THE ECONOMICS OF
RESIDENTIAL LOCATION

The Economics of Residential Location

ALAN W. EVANS
Centre for Environmental Studies

Macmillan

First published 1973 by
THE MACMILLAN PRESS LTD
London and Basingstoke
Associated companies in New York
Dublin Melbourne Johannesburg and Madras

SBN 333 14847 9

Printed in Great Britain by
R. AND R. CLARK LTD
Edinburgh

TO MY PARENTS

Contents

Acknowledgements

The author and publishers are grateful to Homer Hoyt Associates for Fig. 8.10, from *Structure and Growth of Residential Neighbourhoods in American Cities* by Homer Hoyt (Washington, D.C., 1968); and to the Institute of Administrative Management for Figs. 11.2 and 11.3, from *Clerical Services Analysis* (1962).

Preface

The theory of residential location set out by William Alonso and others in the early sixties was an excellent, if complex, piece of theoretical analysis. However, very little empirical evidence was produced to confirm the theory directly. There were two consequences: the excellence (and the complexity!) of the theory established the respectability of urban economics as a subject suitable for study by economic theorists, while the complexity and the lack of evidence led the more empirically minded economists to express doubts as to the theory's validity, ranging from Britton Harris's tentative agnosticism to Harry Richardson's downright atheism, and to its neglect by geographers, sociologists, and town planners.

My aim in this book is to remedy the latter situation by presenting the economic theory of residential location clearly and simply, wherever possible proving results diagrammatically. I have brought out at each stage the empirical evidence on which the theory can be based, and the facts which the theory can explain. Also, I introduce additional assumptions, derive predictions from the theory, and present empirical evidence which generally confirms these predictions. I believe that, as a result, the theory of residential location is now more highly confirmed than most other economic theories.

I started work on this subject as a postgraduate student, initially while holding an exchange scholarship at the University of Michigan in 1965, then at University College London between 1965 and 1967, my Ph.D. thesis being finally submitted to the University of London in 1972. This book is a slightly shortened version of that thesis, the main omission being a chapter, which has been published elsewhere, on the theory of the valuation of time.

I owe a great debt to my supervisor, Professor Marian Bowley. There are certain to be errors of fact and logic in this book but had I not had the benefit of Professor Bowley's keen criticism there would have been many more. I soon found that

I could never get away with anything; that weak arguments would be spotted, immediately exposed, and rejected. I apologise to the reader for those errors which remain: they are my fault alone, for no one could have done more than Professor Bowley to save me from myself.

Many people have argued with me about the theory of residential location, and where possible I have specifically acknowledged my debt to them in the text or in a footnote. I thank them for making me clarify my ideas and arguments. Material from this book has been used in lectures at University College, London, and at the University of Glasgow. I, for one, enjoyed and benefited from the discussions which followed. The comments and arguments of my colleagues and friends at the University of Glasgow were extremely helpful; my thanks are particularly due to Sister Annette (Anne Buttimer), Gordon Cameron, Diane Dawson, Michael Farbman, William Lever, Peter Norman, and Allan Sleeman. At the Centre for Environmental Studies, Michael Ball, David Donnison, and Alex Henney commented helpfully on drafts of the book. I have also benefited from comments by Harry Richardson (University of Kent at Canterbury), Robert Solow (Massachusetts Institute of Technology), and the late John Lansing (University of Michigan).

I am indebted to my wife, Jill, to Mrs Hilda Walker at the University of Glasgow, and to Mrs Celine Gunawardene at the Centre for Environmental Studies for typing various drafts of this book.

Finally, I must apologise to my wife and thank her for her forbearance when I have been working on this book. Many times in the last seven years she has realised from my distant expression that though I was with her in body, in spirit I was locating a residence somewhere.

ALAN W. EVANS

Centre for Environmental Studies,
London
December 1972

1 Introduction: Theories of Residential Location

Until recently, urban problems in Britain have generally been neglected by economists. On the face of it this is odd, for one would expect that the twentieth-century enthusiasm for town planning would have acted as a catalyst for studies in urban economics. Little resulted, however, largely because the view of the British town planner was that urban problems arose through the unfettered operation of the market and that they would be solved when the allocation of land uses was completely in the hands of the town planner. In the view of Sir Patrick Abercrombie, 'mankind might well be divided into two groups, in regard to their surroundings: those who instinctively set about shaping their environment and those who are content to accept the state of things as it exists' (Abercrombie, 1959, p. 9). Economists are put firmly in the second group, town planners in the first. Abercrombie regards the economist as a 'muddler who will talk about the Law of Supply and Demand and the liberty of the individual' (p. 27). Urban problems, then, were seen as planning problems, and the economist was not thought to have any useful contributions to make in the search for solutions.

In the United States, on the other hand, because of the differing political and cultural heritage the British system of 'enforced' planning was largely impossible. The task of the American city planner was to predict the future pattern of land use rather than to lay down edicts as to what that pattern must be. It was generally accepted that the future uses of privately owned land would be determined by the operation of the market and that the planner would decide the optimal distribution of public investment. This approach has resulted in the development of methods of prediction and necessitated the involvement in the planning process of economists and other social scientists.

The work done by R. M. Haig (1926) for the 1928 regional plan of the New York metropolitan region is an example of this involvement but, despite his early work, little further research was done until impetus was given to the development of predictive techniques by concern over the urban traffic problem. This concern led to the development of survey methods, mathematical models, and computer techniques, with the aim of forecasting both the distribution of traffic flows at some future date and the effects of different patterns of road construction. The work of Mitchell and Rapkin (1954) brought out the relationship between patterns of land use and patterns of traffic flow and showed that the traffic flows generated at a site could be predicted if the nature and intensity of land use at the site were known.† Thus, the ability to predict the future patterns of land use grew to be of crucial importance to the transport planning agency.

These developments in American city planning had their effect on British town planning, particularly through the adoption of American transportation study methods in an attempt to solve the British traffic problem. In both countries, therefore, there is now more interest than formerly in the investigation of urban problems by economists.

ECONOMISTS AND RESIDENTIAL LOCATION –
BEFORE 1960

This interest in urban economics has resulted in economists paying attention to the location of activities within the city and, hence, to the growth of interest in the economics of residential location, but prior to 1960, residential location as a field of study was left to sociologists and geographers. The location of the household was seen by economists as resulting from non-economic social factors. Thus Isard (1956, p. 144) stated that 'if consumers are households, we are not able thus far to account for the transport inputs for which they are actively responsible'; he adds in a footnote that

† A concise summary of the history of American city planning, elucidating the close relationship between city planning and transport planning, is given in Fagin (1967).

to do so would take us into the realms of sociology and social psychology. For, to explain the spatial distribution of household consumers around focal points – for example, the population spread around any given metropolitan core – requires knowledge of the process by which tastes are molded and, in particular, understanding of the space preferences of consumers.

Lösch (1954) saw the individual's choice of residential location as a choice between locations in different towns (for work at different workplaces) rather than a choice between locations in a single town. He argues that the occupants of a particular place will be 'all those who achieve their highest utility there' (p. 246). In his view, the utility arising from location at any place will depend on the money wage obtainable there, the prices of goods at that location, and

> those individual characteristics of places and people that cannot be interchanged, all those imponderables of production and consumption that often mean incomparably more to us than the economic process proper. . . .
>
> Total utility . . . differs interlocally for individuals by far more than traveling expenses, for it makes an enormous difference whether we were born in a place or have to move there. Migration means relinquishing much that, like friends, can be replaced only after a long time; or never, like one's native place. We cannot take landscape and people with us (p. 242).

It follows that, in Lösch's view, a person will change the place of his residence (and the place of his work) only if he is induced to do so by the offer of a higher real wage to compensate him both for the loss of a home and for his travelling and removal expense. Since 'the same wage and other identical circumstances provide a newcomer with less utility than they give to a native with the same characteristics' (p. 243), it is implied that households will not be very mobile and, hence, that there cannot be any comprehensive theory, economic or otherwise, which explains the patterns of residential location which may be observed, and that these patterns must therefore have arisen by chance. Further backing for the negative view that economics could not be used to explain the spatial pattern of urban land uses could also be found in Turvey's study of the economics of real property, though it is now clear that his argument was misinterpreted. Turvey stated that:

> If conditions were different and buildings had very short lives, the actual shape and form of a town would be close to its equilibrium pattern. . . . But since this is not the case, since most towns are not in equilibrium, it is impossible to present a comparative statics analysis which will explain the layout of towns and the pattern of buildings; the determining background conditions are insufficiently stationary in relation to the durability of buildings. In other words, each town must be examined separately and historically (Turvey, 1957, pp. 47 f).

Alonso takes this to mean that an equilibrium model of urban land uses would be worth little because, due to the imperfections of the land market, institutional factors, and the permanence of the buildings and structures 'the value of land is of minor importance'.† But, in the passage quoted, Turvey is talking only of the supply side, of the actual stock of buildings which exist at any one time in the city. In other passages he implies that changes in the uses of the buildings (i.e. on the demand side) can take place much more easily.

> Given the existing standing stock of buildings, the pattern of rents can be analysed by examining demand conditions. . . . The point is that the actual rent and use pattern of a given stock of buildings will approach the equilibrium pattern within a period sufficiently short for changes in that stock . . . to be negligibly small (Turvey, 1957, p. 48).

Thus, the buildings on a site may be converted to suit the activity which will pay the highest rent for their use, e.g. houses may be converted into offices or split up into flats.

This interpretation of Turvey's views is in accord with a later statement of his that an analysis of the demand for urban land uses in the long run 'could equally well be regarded as relating to the short run, explaining the equilibrium allocation of a given amount of floor space in a stock of existing buildings' (Turvey, 1964, p. 845).

† Alonso (1964a, p. 12). See also Kirwan (1966).

THE HISTORICAL THEORY OF RESIDENTIAL LOCATION

Although economists may have felt that they had nothing to contribute to the theory of residential location, the theory used by sociologists and city planners to explain patterns of location was clearly economic since it depended on the assumption that, as the population of a city grew and the housing stock increased, the newest dwellings would always be occupied by the highest-income groups and that, as the dwellings aged, they would filter down through the population, becoming cheaper and cheaper and being occupied successively by households of lower and lower incomes. The pattern of location of households with differing incomes would be determined by the pattern of growth of the city in the past. For this reason Alonso (1964b) described this type of theory as an historical theory.

The earliest version of this theory was put forward by E. W. Burgess in 1925 as part of a general theory of the growth of the city and has come to be known as 'the concentric zone theory'. In Burgess's view,

> an ideal construction of the tendencies of any town or city to expand radially from its central business district [would show that] encircling the downtown area there is normally an area in transition which is being invaded by business and light manufacture. A third area is inhabited by the workers in industries who have escaped from the area of deterioration but who desire to live within easy access of their work. Beyond this zone is the 'residential area' of high-class apartment buildings or of exclusive 'restricted' districts of single family dwellings. Still farther, out beyond the city limits, is the commuters' zone – suburban areas, or satellite cities – within a thirty- to sixty-minute ride of the central business district (Burgess, 1925, p. 50, references to a diagram omitted).

Thus, the city is represented in the theory as a series of concentric zones, though 'it hardly needs to be added that neither Chicago nor any other city fits perfectly into this ideal scheme' (pp. 51–2). Burgess argues that the main result of the growth of the population of the city is an expansion in its area and a consequent 'tendency for each inner zone to extend its area by the invasion of the next outer zone' (p. 50), thus resulting in a particular site serving a succession of users. The reasons for the concentric ordering of the zones are not made clear, though it is implied that, because of the growth of the city, the oldest residential property will tend to be near the centre (i.e. in the zone in transition) and the newest property will tend to be in the commuters' zone beyond the city limits. The highest-income groups are able to afford the newest housing and so locate beyond the city limits. The poorest are only able to afford the oldest property which will be in the zone in transition.

A second version of the historical theory called 'the sector theory' was published by Homer Hoyt in 1939. He, too, argues that as 'residential structures deteriorate and become obsolete with the passage of time', they are occupied by successive groups of people with lower incomes and lower social standards (p. 111) but, on the basis of his study of property inventories of 142 American cities, he argued that the pattern of residential location could be explained in terms of sectors.

The high-rent area, he stated, is initially close to the retail and office centre on the side of the city farthest from the manufacturing district, but growth in the population and area of the city leads to a migration of the high-rent sector radially outward from the centre. Movement in any other direction is impossible since

> the wealthy seldom reverse their steps and move backward into the obsolete houses which they are giving up. On each side of them is usually an intermediate rental area, so they cannot move sideways. As they represent the highest income group, there are no houses above them abandoned by another group. They must build new houses on vacant land. Usually this vacant land lies available just ahead of the line of march of the area because, anticipating the trend of fashionable growth, land promoters have either restricted it to high grade use or speculators have placed a value on the land that is too high for the low-rent or intermediate rental group. Hence the natural trend of the high-rent area is outward, toward the periphery of the city in the very sector in which the high-rent area started (Hoyt, 1939, p. 116).

The direction of movement of the high-rent area, and the sector of the city in which it lies, 'is in a certain sense the most important because it tends to pull the growth of the entire city in the same direction'. This is because 'lesser income groups seek to get as close to it as possible'. There will therefore be new building for these lesser-income groups in the adjacent sectors and for the higher-income group 'on the outward edges of the high-rent area. As these areas grow outward, the lower and intermediate rental groups filter into the homes given up by the higher income groups' (p. 114).

According to the sector theory, the pattern of residential location is not completely explained by the filtering-down process. The higher-income groups will occupy new houses in the best residential land in the high-rent sector. The lesser-income groups also can afford new housing and therefore will locate in the adjacent sectors. 'Occupants of houses in the low-rent categories tend to move out in bands from the centre of the city mainly by filtering into houses left behind by the high-income groups or by erecting shacks on the periphery of the city' (p. 120).

Hoyt also specifies the factors which seemed to have governed the direction and pattern of growth of the high-rent areas in the cities which he studied. These factors are variously topographical, sociological, historical, etc. Thus the high-rent areas are said to grow along the fastest transport routes, towards an existing built-up area, towards the homes of the community leaders, towards high ground, along lake fronts, or towards open country. In addition, Hoyt notes that 'de luxe high-rent apartment areas tend to be established near the business center in old residential areas. . . . When the high-rent single-family home areas have moved far out on the periphery of the city, some wealthy families desire to live in a colony of luxurious apartments close to the business center' (p. 118).

'TRADE-OFF' THEORIES OF RESIDENTIAL LOCATION

When economists became interested in the location of activities within the city during the late fifties, the problem of residential location was approached in a different way. Instead of regarding the location of a household as being determined by the availability of housing, the household was assumed to find its optimal location relative to the centre of the city by trading off travel costs, which increase with distance from the centre, against housing costs, which decrease with distance from the centre, and locating at the point at which total costs are minimised. Richardson (1971) describes this type of theory as a trade-off theory, but it has also been called the least-cost theory of residential location by Beed (1970),[†] who notes that the theory was first formulated by the sociologist L. F. Schnore in 1954. Schnore assumes that 'rent, or the cost of occupancy of a site, *declines with distance* from an activity center' but 'transport costs are assumed to *increase with distance*' so that 'the maximum distance from significant centers of activity at which a unit tends to locate is fixed at that point beyond which further savings in rent are insufficient to cover the added costs of transportation to these centers' (p. 342).

The first economists to explain the pattern of residential location in a city by the trade-off between transport costs and housing costs were Hoover and Vernon (1959) who, in their study of the New York region, represented the household's choice of location as its solution of the conflict between spacious living and easy access. More rigorous versions of the theory came with the work of William Alonso (1960, 1964a) and Lowdon Wingo Jr. (1961a, 1961b).[‡]

Both versions were developed to explain the relationship between land values, or rents, and accessibility so that instead of assuming that land values declined with distance from the centre this was a conclusion derived from the theoretical analysis. Neither theorist was primarily concerned with patterns of residential location; Alonso set out to construct a general theory of land rent, and Wingo wished to investigate the effect of changes in the transport system on urban land use. However, both studies present rigorous theoretical analyses of an urban spatial economy in which it is shown how, under conditions of perfect competition, the household chooses its location to maximise its utility, and in doing so balances 'the costs and

[†] It is called the structural theory by Alonso (1964b) on the grounds that it represents the working out of tastes, costs, and income in the structure of the market but this name is, I think, misleading since it could as well designate the historical theory on the grounds that the pattern of location is determined by the city's existing structure.

[‡] It may be noted that another version of the theory by Kain (1962) appeared at about the same time.

bother of commuting against the advantages of cheaper land with increasing distance from the center of the city and the satisfaction of more space for living' (Alonso, 1960, p. 154).

It is interesting to note that, while the historical theory explains the pattern of location solely in terms of the supply of space – namely, the age and location of the stock of housing – the versions of the trade-off theory presented by Wingo and Alonso explain the pattern of location solely in terms of the demand for space. In their theories, the buildings of the city are ignored completely and the household is concerned only with the quantity of land it occupies and its location. This rather extreme assumption is modified by Richard Muth who incorporates a theory of the supply of space into the version of the trade-off theory set out in his study of the operation of the urban housing market published in 1969.†

AN OUTLINE OF THE THEORY OF RESIDENTIAL LOCATION

In this book we start with the simplest possible assumptions and gradually make them more complex. At each stage in the construction of the theory, additional predictions and explanations are obtained. The danger with this approach is that the reader may be dissatisfied with the early analyses, when the simplest assumptions are made, feeling that these analyses ignore what he regards as the more important determinants of patterns of residential location, and this hostility may be carried over into his reading of the later chapters of the book when the theory becomes more complex and, hopefully, more realistic. In this section we therefore outline the model so that the reader will be aware that assumptions which he may object to in the earlier chapters may be altered later.

As we have mentioned, none of the economists who developed the theory of residential location has been directly concerned with residential location. In our view, the theory has therefore not yet been fully developed; it has appeared unrealistic because only the essentials of the theory have been developed, and the explanation and prediction of which the theory is capable have not been seen. Left as a simple piece of theoretical

† The book, however, was mainly written between 1959 and 1964.

analysis the theory is clearly unsatisfactory, and many authors have consequently cast doubt on its usefulness or relevance.† This is understandable if the theory is considered in its undeveloped state, but it is a situation which we hope to remedy in this book, by introducing new assumptions into the theory and presenting the empirical evidence which tends to support the theory.

Initially, we assume that the workplace of all the workers in the city is at a single central workplace (the central business district or CBD), and that the cost and speed of travel is a function of distance from the CBD. If the only locational variables taken into account by the householder are the time spent travelling, the cost of travel, and the quantity of space occupied, then, if he either maximises utility or minimises his total rent plus travel costs, the value of a standard unit of housing space (called a space unit) must decline at a diminishing rate with distance from the centre. In Chapter 4 we introduce a theory of the supply of space and demonstrate that, if developers and landlords maximise profits and if housing space is assumed to last indefinitely, the value of land and the density of space units must also decline at a diminishing rate with distance from the CBD. In Chapter 5 we show that population density will probably also decline at a diminishing rate with distance from the CBD, and we bring together the empirical evidence which confirms the various predictions about the rate of decline of rents or property values, land values, and densities.

In Chapter 6 we show that, on the basis of the assumptions made, equilibrium in the market for space is attainable in the city, given the size and characteristics of its population. Up to this point we have refined the trade-off theory, strengthening its theoretical base and presenting it diagrammatically to make it more easily understandable. In the first part of the chapter we assume that people are not indifferent to the density of the area in which they live but prefer lower densities to higher, and show that it may then be necessary to control densities by some form of government intervention. The results show that we can draw no normative conclusion from our theory of residential

† See, for example, Harris (1968), Johnston (1971), Richardson (1971), and Timms (1971).

location, even though the pattern of location may not be affected in a predictable way.

In the second part of Chapter 7 we drop the assumption that housing space can be assumed to last indefinitely. Using two types of model we show, however, that this is unlikely to have an important influence on patterns of residential location; the method of renting or owning property seems irrelevant, and the quality of housing seems, with qualifications, to depend on the characteristics of the occupier and the immediate neighbourhood, rather than vice versa. This means that we can discount the importance of filtering down and the associated historical theories of residential location which are based on the assumption that housing filters down from one income group to another as it ages and its quality deteriorates.

Accordingly, in Chapter 8, we use the trade-off theory to explain the relative location of different income groups in the city. We show that the pattern of location will depend on the dispersion of the possible range of income elasticities of demand for housing among the population, and on the distribution of incomes in the population, relative to the cost, speed, and comfort of travel. Under plausible conditions we show that a likely pattern is one in which the rich live at the centre of the city, the very poor live adjacent to them, and incomes generally then increase with distance from the centre so that the very rich live also at the edge of the city.

At this point we introduce the assumption that households in a given socioeconomic group will prefer to locate in the same neighbourhood as others in the same group, both because they will tend to draw their friends from this group and because the goods and services they require will tend to be provided by traders in this area. Therefore, instead of households in different socioeconomic groups locating in concentric zones about the CBD, they will tend to group together in neighbourhoods and this will result in the existence of high and low-income sectors.

In the final section of Chapter 8 we investigate the process by which the high-income neighbourhood in the inner city may expand in area when the number of people wishing to live there increases. We show that it can lead to the upgrading and renovation of housing previously occupied by low-income households, but that the high-income people who first move into the low-income areas are likely to be people for whom the external effects of a low-prestige address are minimal, people who are not in business, but in teaching, architecture, or similar professions.

In Chapter 9 we turn from economic status to family characteristics and show that since a decrease (increase) in the proportion of the household working leads to a reduction (increase) in the importance of travelling costs relative to space costs, households with a low proportion working are likely to live further from the CBD than those with a high proportion working. Thus, households with children of school age or below and a wife not in paid employment are likely to live further from the centre than households with no children, where the wife is likely to be in paid employment and the demand for space is likely to be less. This analysis is confirmed by the empirical evidence.

In Chapter 10 we conclude the analysis of patterns of location in a monocentric city by using comparative statics to show that the effect of an improvement in transport speed is likely to be a reduction in the slope of the rent, density, and land-value gradients, a 'flight to the suburbs' by high-income households, and possibly, the existence of an inner ring of old, high-density housing which it is unprofitable to demolish.

In Chapter 11 we assume that as well as the CBD there are a number of subcentral workplaces in the city, the number and characteristics, of the workers at each being given. We show that, in equilibrium, wage rates at each subcentre should be inversely correlated with its distance from the CBD, but that, although there is clear evidence that this wage gradient exists in London, there is no evidence that it exists in other cities. In the two following chapters, however, in which we investigate the pattern of the journey to work in a multinuclear city, we assume, in order to ensure equilibrium, that wage rates do vary within the city. We then show, in chapter 12, that workers in subcentres are likely to live in the same sector of the city as their place of work is, and conversely, that residents in a particular suburb are most likely to work at a workplace located between their place of residence and the city centre. In the final section of the chapter we show that the deterministic theory so far considered could be turned into a stochastic model of the journey to work for operational purposes.

In Chapter 13 we show that a sectoral pattern of journeys to work may be present even when the place of work is a part of the CBD of a large city, and we present an econometric model of the journey to work in the City of London, a part of the London CBD which shows that direction, distance, and the socioeconomic status of the inhabitants are the main factors determining the number commuting to the City of London from any residential area in the rest of London.

In Chapter 14 we use the results set out separately in each of the later chapters to explain the pattern of variation of household size in the city. In doing so we integrate the analysis of the pattern of location of households of differing size in Chapter 9, the analysis of the pattern of location of households with differing incomes in Chapter 8, and the analysis in Chapters 12 and 13 of the pattern of the journey to work in a multi-nuclear city.

Finally, in Chapter 15, we give several illustrations of the uses of the theory of residential location in understanding the consequences of urban plans and policies.

2 The Theory of Residential Location: Limitations, Assumptions, Definitions

In this chapter we cover some of the preliminary work necessary for the development of the theory. In the first part of the chapter we state some of the limitations of the theory; the rest of the chapter is used to state assumptions and to define terms. For some of the assumptions it is argued that they are at least an approximation to reality. For others, it is argued that they may be more realistic in the case of some cities than others but that it is necessary to make the assumptions explicit to understand both the limitations of the theory, and if possible, the way in which the predictions of the theory would change if different assumptions were made.

THE LOCATION OF THE INDIVIDUAL HOUSEHOLD

It should not be thought that the theory can be used to explain the location of any particular household. It can only explain the pattern of location of households in general. This is not unusual in economics. Hicks's discussion of this point in the introduction to his theory of economic history is worth quoting at length, for he states the position well.

> It will be the group, not the individual, on which we shall fix our attention; it will be the average, or norm, of the group which is what we shall be trying to explain. We shall be able to allow that the individual may diverge from the norm without being deterred from the recognition of a statistical uniformity. This is what we do, almost all the time, in economics. We do not claim, in our demand theory for instance, to be able to say anything useful about the behaviour of a particular consumer, which may be dominated

by motives quite peculiar to himself; but we do claim to be able to say something about the behaviour of the whole market – of the whole group, that is, of the consumers of a particular product. We can do this, it must be emphasised, without implying any 'determinism'; we make no question that each of the consumers, as an individual, is perfectly free to choose. Economics is rather specially concerned with such 'statistical' behaviour (Hicks, 1969, p. 3 f.).

CITY SIZE

The basic assumption of the trade-off theory is that the householder attempts to minimise the cost of his location by trading-off rents against travelling costs and that the primary factor in determining his location is the total rent plus travelling costs which he has to pay at any point. These costs vary with distance from the place of work. When all the households in the city attempt to minimise costs in this way the result is a definite (predictable) pattern of location of the different types of households. Those for whom proximity to the place of work is most valuable locate near it. Those for whom proximity to the place of work is least valuable locate on the periphery of the city. For this pattern to result it is, however, necessary that a substantial difference exists between the rents or travelling costs at locations near the centre and those on the periphery. In a large city, simply because of the large area which it covers and the long distances which have to be travelled, this will be true but it need not be true in the village, in the town, and possibly even in the small city.

An example will illustrate the point. In London, in 1967, the distance from the centre to the edge of the city was approximately 12 miles. The travelling time for the journey by underground would be about 40 minutes for the one-way journey, and the financial cost would be about 17½p. Over the week, therefore, the cost of the journey to work in the centre would be about £1·75 and the time spent on the journey would amount to nearly seven hours. The total cost to the individual of this long journey to work would therefore be very high. On the other hand, the savings in housing costs which were possible were equally high. In the same year the price of a two-bedroom flat on a 99-year lease varied from approximately £5,000 on the periphery to £15,000, or more, near the centre. With such tremendous variations in housing and travelling costs within the built-up area it is likely that the individual will be led to choose his residential location carefully, and that his choice of site will be dictated mainly by the relationship between his total housing and travel costs.

The smaller the city, however, the smaller this variation in costs with distance becomes, since the radius of the city becomes correspondingly smaller. The householder may therefore ignore the relative location of a site when the travelling costs fall very low. For example, if the radius of the town is only a mile the direct transport costs will amount to only a few pence per week however far the householder travels within the built-up area, and correspondingly, the time spent on the journey to work will be negligible. If he can ignore these costs it is likely that the householder's choice of site will be decided by other factors. For example, he may take into account topographical features such as hills, lakes, rivers – as suggested by Hoyt – and give these factors more weight than he would if he lived in a larger city. Or the condition of the housing available may be the determining factor, as suggested by Burgess. For these reasons the pattern of residential location in a small city is likely to be more complex than in a large one, and less amenable to analysis by deductive reasoning.

The point here was made more generally and in a different form by Haig, in 1926. He argued then that New York was a peculiarly good place to study the problem of intra-urban location.

> Its size and complexity, which at first glance appear to be such serious obstacles, prove upon examination to carry with them great advantages. The magnitude of the metropolis not only minimises the influence of 'sport' cases, but it also operates, like a Bunsen flame under a test tube, to produce phenomena which do not become explicit in small places where the pressure for space is not great. Again, the complexity of New York implies a high degree of segregation of economic activities, and consequently an opportunity to

observe and to distinguish among many rather than a few economic functions struggling for the more convenient locations (Haig, 1926, p. 407.).

Haig is here arguing that a large city is a good place to study location theory inductively. The point made in this section is, in a sense, the converse of this. A deductive location theory is a poor predictor of patterns in small cities.

It should also be noted that the argument applies to any small area in a large city. We do not attempt to explain the pattern of location within a small area. This pattern will be determined by factors other than those considered in the theory because the cost of work travel from one end of the area to another will be small, and the variation in other variables will be more important. Furthermore, to explain the pattern of location in a small area, we would have to come close to explaining the location of individual households, and as we stated in the first section of the chapter, this is impossible.

RESIDENTIAL DENSITY AND EXTERNALITIES

This book is an essay in positive urban economics. Our aim 'is to provide a system of generalizations that can be used to make correct predictions about the consequences of any change in circumstances' (Friedman, 1953), as for example, in Chapter 10, we use the theory of residential location to predict the effects of changes in the transport system. In general we are not concerned with the welfare economics of residential location; we consider only what the patterns of location are, not what they ought to be.

It is particularly important to note that we ignore two sources of external effects because, though they may affect welfare, we do not think that they cause important systematic variations in the pattern of residential location. In the first place we ignore the external effects of things peculiar to individual cities and locations. A factory or a park may distort the pattern of residential location in a city – because of the fumes from the factory, or the beauty of the park – but this distortion cannot be handled in a theory which aims to be generally applicable, since it is peculiar to a particular part of a particular city.†

Secondly, and possibly more important, we ignore the external effects of variations in residential density.‡ We assume that the householder is indifferent to differences in residential density; he demands a certain amount of space and does not mind whether this is provided at a high density in blocks of flats or at a low density in suburban housing. This is, of course, a very questionable assumption, but in Chapter 7 we try to show that, even if we assume that householders prefer lower densities, the patterns of location are unlikely to be different. As we show there, these preferences will not usually be reflected in market transactions because the density at which each individual lives depends on the actions of other people, and the degree of co-operation necessary to live at lower densities may not be possible. In explaining and predicting patterns of location we therefore ignore density, but this does not mean that the patterns of location which we predict are optimal. Because co-operation to ensure lower densities is difficult, densities may be higher than people would wish. An optimal pattern of location may require that densities be lower throughout the city; moreover, it may be that the divergence from the optimal density is greatest in the inner city where densities are highest.

Thus, the economic welfare of the inhabitants of the inner city may be reduced relative to that of the inhabitants of the outer areas where densities are lower and the divergence from the optimal density is least. If, as we show in Chapter 8, the inner city is mainly occupied by lower-income groups who are precluded from living further out by the cost of travel, the divergences from optimal density of the inner areas may result in a systematic reduction in the economic welfare

† It has also been suggested by Crecine, Davis, and Jackson (1967) that these external effects are unimportant. After a study of property transactions in Pittsburgh in the period 1953 to 1963 they concluded that 'the results . . . do not support the notion that external diseconomies (or external economies) abound in the urban property market. Indeed, these results suggest that there is a great deal of independence in that market' (p. 93).

‡ I am indebted to Marian Bowley for persuading me of the importance of this.

of the lower-income groups relative to the upper-income groups.†

THE PHYSICAL ENVIRONMENT

We generally assume in this book that the city is located on a flat, featureless plain no part of which differs from any other part. The environment, as such, is scarcely mentioned and the quality of the environment is never explicitly assumed to be a variable which is taken into account by consumers in deciding their optimal location.

On the face of it this might seem an important omission; Richardson (1971) has argued that the quality of the environment is one of the most important determinants of a householder's location. In our view, the omission is more apparent than real. The quality of the environment is a somewhat amorphous concept, but its variation in an urban area would seem to depend mainly on variations in residential density, variations in average socioeconomic status, and lastly, variations in the terrain. At least one author has, in fact, equated the quality of the environment with residential density (Mirrlees, 1972). We do not wish to make this extreme assumption; we merely note that in Chapter 7 we do discuss the economic implications of consumers preferring lower densities to higher and that those who wish to read this as a discussion of the economic implications of a preference for better environments may do so.

In our view, the most important determinant of the quality of the environment in an area is the socioeconomic status of those living there. It is probable that environmental quality is a superior good and, hence, that higher-income people will spend proportionately more on the area over which they have control than will lower-income people (or their landlords). The external effects of this expenditure will improve the environment of those living in the neighbourhood, who will tend to be of the same socioeconomic status. Furthermore, those with

† This systematic distortion of the distribution of economic welfare may be all the greater if, as we show in Chapter 10, past changes in transport technology have made densities in the inner areas higher than they otherwise would be.

higher incomes will consume more space at any given price and will live at lower residential densities than those with lower incomes. They will thus have greater control over their immediate environment and a greater incentive to improve it, since the external effects of their expenditure will be less in comparison with the internal effects. Therefore social agglomeration, a concept which we introduce in Chapter 8, both explains and determines variations in the quality of the environment in urban areas.

Finally, it should be noted that, in Chapter 8, we argue that the areas of high economic status may be partially determined by topographical features such as hills, rivers, lakes, etc. since people with high incomes may be willing to pay more than others to live in areas with good natural features. In this way, but with the qualification made earlier in this chapter that these topographical features will be more important in small towns than in large cities, we do consider the effect of variations in the terrain.

While we do not, therefore, explicitly study the quality of the environment, we do study separately its three most important determinants, residential density, social agglomeration, and the terrain.

HOUSEHOLD MOBILITY

The patterns of location which we predict are the patterns which would exist when the urban system is in equilibrium. In effect, it is assumed that the system can attain equilibrium and, hence, that there are no restrictions on household mobility and that households are able and willing to change their location should the circumstances which determined their original location change.

In the real world there are, of course, considerable barriers to movement. The owner-occupier must incur legal and other charges both in selling a house and buying a new one. These once-for-all costs are a disincentive to mobility. The barriers to movement are less in the case of the household occupying privately rented accommodation, but the operation of laws controlling rents may restrict mobility. Often the rent of the dwelling occupied by the household may be controlled, and a move would mean that the household would have to pay a far higher rent. The barriers to movement may be greatest for tenants of local-authority housing. The household is allocated a dwelling at a subsidised rental. The location of this dwelling may not be optimal for the household but may be acceptable because of the subsidy. Hence, as we show empirically in Chapter 14, the pattern of residential location in a city in respect of the occupiers of local-authority housing will be that decided by the local authority, and not that which would exist were each household to be given a housing subsidy to be spent on housing at the location desired.

Moreover, the various barriers to mobility mean that the urban system is usually in a state of disequilibrium, though moving slowly towards some equilibrium. In these circumstances one constructs theories which explain the situation in equilibrium, first, in the hope that the system is not too far from equilibrium, and second, as a step towards the construction of more-complicated theories which explain the changes taking place when the system is in disequilibrium.

THE CONCEPT OF THE SPACE UNIT

The argument of this section is an attempt to settle a methodological problem. In the construction of the theory it is assumed that it is possible to divide all residential living space into 'space units'. A space unit may be a unit of area of garden or other open space, or it may be a unit of area of floor space. A dwelling (including its garden) constitutes a definite number of space units. Just as the interior of a dwelling comprises a definite area in square feet of floor space (or volume in cubic feet), and the garden may be said to cover a definite number of square feet of ground, so the dwelling and garden together may be said to be a definite number of space units.

This concept allows one to express a person's demand for residential living space in terms of space units, to express the supply of space by the developer of an area of land in terms of space units, and thus to use a variation of supply and demand analysis in the theory. The 'space unit' is equivalent to the 'unit of housing services' which Muth uses in his version of the trade-off theory.

There would seem to be two alternatives to using this kind of concept. The first alternative is to assume that the demand for residential living space can be expressed in terms of a demand for an area of ground alone. This is the procedure adopted by Alonso and Wingo.† It is, however, an awkward assumption since it implies that the inhabitants of a block of flats demand, and occupy, only a small fraction of the ground area of the block. This is an unrealistic assumption since it is obvious that the occupier of one of the flats derives satisfaction from, and therefore demands, the living space in his flat and not the small area of ground which is its share of the ground area of the block.

A further implication of this alternative assumption is that the quantity of residential living space on a particular area of land is fixed and cannot be changed. But the supply of space units is not fixed, since it is always possible that one more floor may be added to a block of flats, thus increasing the supply of space units on that area of ground. The result of increasing the height of the block may be that another family is added to those living in the block. It would lead to unnecessary complications if ground area alone were considered, since the ground area per family would diminish. If the construction of the new floor was caused by improvements in building technology, it would be nonsense to describe the families' demands for space as having fallen, yet this is the only way of describing the change if the demand for space is expressed in terms of ground area. The use of the concept of space units allows the construction of a supply curve of space units per acre, and this can be used to investigate the effects of changes in building technology on the spatial distribution of the residents of the city.

The second alternative to the use of the concept of the space unit is to assume that the demand for space can be split up into the demand for garden or open space and the demand for living space. In the author's opinion, this would unnecessarily complicate the theory without improving either its realism or its predictive power. As will be shown later, the supply of space units can anyway be represented as a function of the value of land, and hence, of the price of garden space, so that the split would be of little use in considering the supply side.

† See Alonso (1964a, p. 17) and Wingo (1961a, pp. 20–1).

A further point against the adoption of this assumption is that it is not necessary that garden or open space should actually be on the ground, since in some areas where land is expensive garden space may be provided in the form of roof gardens or balconies. The distinction between garden (ground) space and indoor (built) space becomes blurred and indefinite when this possibility is considered. The use of two concepts of space therefore solves fewer problems than it creates.

The more that the alternatives are considered, the more it seems that the use of the concept of a space unit creates the fewest difficulties. Nevertheless, the objection may be made that the concept is inadmissible if it cannot be defined in terms of floor space or ground area. It may be felt that one should be able to define one space unit as equal to x ft^2 of floor space or y ft^2 of ground area. This cannot be done. The space unit exists as an entity only in the theory and is a theoretical construct which can be partially interpreted but cannot be explicitly defined. The use of a theoretical construct is not unknown in economics, indeed utility itself is a theoretical construct, but its use occurs most often in the physical sciences. Illustrations of the use of theoretical constructs therefore most often give examples from physics. For example,

> no direct interpretation for 'mass of a gas molecule' and '$\bar{v}$' can be given. To define 'v' in terms of 'average displacement per second of a molecule' is not to interpret it, for there is no such thing as a direct measurement of displacement of a single gas molecule in random motion. . . . The constructs of the kinetic theory [of gases] are indirectly but only partially interpreted: experimental data, such as results of measurement of gas pressure and gas temperature, will be relevant to the question whether the postulates containing these constructs are true, and in this sense the constructs are empirically significant (Pap, 1963, p. 53f.).

An illuminating analogy for a theory which illustrates the role which theoretical constructs play is that of a net:

> the empirical significance of a scientific term depends on the theoretical network in which it is embedded. A partially interpreted deductive theory . . . is comparable to a net only some of whose knots are anchored to the ground by posts;

these knots correspond to the theoretical terms that are directly interpreted, the posts to the interpretative sentences (variously called 'rules of correspondence', 'coordinative definitions', 'reduction sentences', 'operational definitions'), the ground to the plane of observation, the freely floating knots to those theoretical terms which owe their empirical significance only to the postulates (including definitions) that connect them with interpreted terms (Pap, 1963, p. 52).

In the case in point, the postulates by which the term 'space unit' is partially defined are: (1) If a house or a flat or other dwelling is identical to a house or flat or other dwelling at another location, then it has exactly the same number of space units. (2) If a dwelling is identical with another dwelling save that the former has a larger garden area than the latter, then the former has more space units. (3) If a dwelling is identical with another dwelling save that the former has more rooms or larger rooms than the latter, then the former has more space units. (4) If a dwelling has more space units than another dwelling, then it has either a larger garden or larger rooms or more rooms. A further postulate is also necessary, stating that an area of floor space is equivalent to a larger number of space units than the same area of land.

The reasons for the fifth postulate require further explanation. If a bungalow is built on an acre of land it is reasonable to suppose that the number of space units on that acre has increased. But if an area of floor space is equal to the same number of space units as the same area of land, the number of space units per acre after the bungalow is built will not have increased, and may even have decreased (because of the thickness of walls, etc.); only if multistorey houses were built on the land would the number of space units to the acre be increased. Intuitively this position is unreasonable, and so the fifth postulate must be laid down in order to avoid it.

Against this postulate it can be argued that a square foot of floor space and a square foot of land differ qualitatively, but not quantitatively, and that this qualitative difference ought not to be concealed by ascribing to them a quantitative difference.†

† This argument seems similar to that directed against the neoclassical theory of production by Joan Robinson (1965).

The two types of space should therefore be considered separately, and the demand and supply for both types of space should be considered as separate markets. The qualitative difference between the two types of space would then be reflected in the relative prices paid for each type of space. This course of action has already been rejected on the grounds that it would unnecessarily complicate the theory, without increasing its explanatory or predictive powers. While therefore acknowledging the force of the argument that the two types of space should be considered separately, the balance of the arguments still suggests that the best solution to the problem is to treat both types of space as divisible into space units. If this is the policy adopted, then the fifth postulate seems necessary to complete the implicit definition of a space unit.

AGE AND QUALITY OF HOUSING

A further assumption is necessary to make the concept of the space unit usable. It is that the number of space units in a dwelling are the same whatever the age of the building. The ageing of a building does not therefore result in a decrease in the number of space units to the acre.

It is recognised that the price the householder is willing to pay per space unit may be less for an old space unit than for a new but, to simplify the development of the theory, it is usually assumed that the space units which are being demanded and supplied are standardised new units, though this assumption is relaxed in Chapters 7, 8, and 10.

The effect of age and obsolescence on the price per space unit is discussed in Chapter 7. The standardisation of the age of a space unit in the rest of the discussion will facilitate the examination of this problem of the age and obsolescence of buildings by isolating it from other factors affecting residential location.

HOURS OF WORK

It is assumed that the worker is free to vary his hours of work. This may be thought to be unrealistic since the hours of work of most jobs are fixed. Nevertheless, in considering the labour market in each city the assumption is not as perverse as it at first seems. There are two arguments in its favour. First, the number of alternative jobs available to a person in a city will be directly proportional to the size of the city. At the city centre the number of alternative jobs available in a very small area is so large that a person seeking a quite rare job may be faced with alternatives.

In a large city, therefore, an employee can choose the job which he prefers, taking into account the rate of pay, hours of work, conditions of work, and other relevant factors. In these circumstances the hours of work are variable in the sense that, though hours of work for any particular job may be fixed, the employee can choose the job with hours of work to his liking. Second, a city is not just one market for jobs at a single place but consists of a variety of labour markets in all parts of the city. The worker therefore has the choice of jobs at various distances from his place of residence. Therefore, if he finds the hours of work at one job too long he may change his job to one at a point nearer to his home, and thus spend less time travelling and have more time available for leisure or work. A worker who finds the hours too short at one place may choose to travel further and try to obtain a job at a higher rate of pay but further from his home.

There is some evidence that hours of work do vary systematically in urban areas. Malamud, in a study of the economics of office location, found that, in the United States,

> hours of work at suburban offices are longer than in downtown offices. Suburban workers likely live near their offices and need not contend with congestion in reaching work. In the New York Metropolitan Area, 36 hours of work are typical for offices in the city proper; 37 hours are typical for Westchester offices; and $39\frac{1}{2}$ hours typical for Nassau County offices. For the Philadelphia SMSA's three inner counties, $38\frac{1}{2}$ hours are typical; $39\frac{1}{2}$ hours are typical for her five outer counties (Malamud, 1971, p. 111).

It therefore seems valid that the labour market is perfect enough for it to be realistic to assume that the hours of work are variable and that the worker chooses his job and hours of work to achieve maximum satisfaction.

PLACE OF WORK

Until Chapter 11 it will be assumed that all firms which employ those resident in a city are located at a single point on the plain. In the final chapters the effects of assuming other workplaces in the city are investigated.

The assumption of a single workplace is not so unrealistic considered in the light of the proportion of the total employment in most major cities which is located at the centre. Table 2.1 sets out the figures for the six British conurbations for which conurbation centres were defined in the 1961 Census. It can be seen from the table that between 10 and 30 per cent of the total employed population is employed in a minute proportion of the total area of the conurbation, giving an indication of the concentration of employment in the centres of the cities. This concentration does not usually occur to the same extent in any other part of the city and it is usual for the density of the working population to decline rapidly with distance from the city centre and to fall to a fairly low and uniform level a few miles from the centre. This level is maintained to the edge of the city save for the presence of a few minor subcentres.

The extent of this decline in employment density is indicated by the figures in the last three columns of Table 2.1 which show the density of employment for the conurbation centre, the central city, and the conurbation (excluding the central city), in respect of each conurbation.

In view of this extreme concentration of employment it seems plausible to approximate it, in the first instance, by assuming all employments to be located at a single point.

TRANSPORT SPEEDS AND COSTS

It is assumed that transport speeds and costs per mile are uniform whatever the direction of travel, though speeds and costs per mile may vary with the distance travelled. Travel to any point one mile from the point of origin of the journey will cost the same, and take the same time, whatever the direction of travel. This is not to say that the cost and speed of travelling a second mile may not differ from the cost and speed of travelling the first, but the cost and speed of travel of the second mile must be the same whatever the direction of travel.

TABLE 2.1

Working Population, Total Areas, and Gross Employment Densities of Six British Conurbations

Conurbation	Number of persons (in thousands) having employment in each subdivision of the conurbation			Area in acres of each subdivision of the conurbation			Employment density (in persons per acre) of each subdivision of the conurbation		
	Conurbation centre	Central city† excluding conurbation centre	Conurbation excluding central city†	Conurbation centre	Central city† excluding conurbation centre	Conurbation excluding central city†	Conurbation centre	Central city† excluding conurbation centre	Conurbation excluding central city†
Greater London	1,414·7	1,175·9	1,897·2	6,683	68,215	386,987	211·7	17·2	4·9
West Midlands	120·9	534·2	567·9	934	50,213	120,860	129·4	10·6	4·7
South-East Lancashire	167·1	261·1	795·6	686	26,569	215,666	243·6	9·6	3·7
Merseyside	157·3	243·6	206·8	937	26,873	68,215	167·9	9·1	3·0
Tyneside	78·3	100·4	207·4	971	10,123	46,618	80·6	9·9	4·4
Clydeside	139·3	363·1	304·9	490	38,157	168,940	284·3	9·5	1·7

Sources: Census 1961, England and Wales: Workplace Tables, Table 1; Age, Marital Condition, and General Tables, Table 3; County Reports, Table 3 (General Register Office, London). Census 1961, Scotland: Occupation, Industry, and Workplace Tables, Part III, Table 1; Age, Marital Condition, and General Tables, Table 3; County Reports, Table 3 (General Register Office, Edinburgh).

† The term 'central city' refers to the local authority area which includes the conurbation centre, viz. the County of London, the County Boroughs of Birmingham, Manchester, Liverpool, and Newcastle upon Tyne, and the County of City of Glasgow, respectively.

This assumption is fairly realistic when the only journey considered is the journey to work and when all workplaces are assumed to be located at the centre of the city. The centre is almost invariably the focal point of a radial transport system which enables journeys in any direction to be made at similar costs and speeds. It is probably unrealistic when off-centre workplaces are considered, but it is a useful first approximation.

PERFECT COMPETITION

It is assumed throughout that those conditions of perfect competition prevail which are usually assumed in economic theory. Some of these conditions have been discussed in detail in previous sections of this chapter. For example, the hours of work are assumed to be variable and the householder is assumed to be free to alter his place of residence. Other assumptions are implied by the postulated conditions of perfect condition. For example, it is assumed that employers and employees have complete knowledge of the alternatives open to them and of the results of their following each course of action. Both employers and employees follow the courses which would lead them to their respective goals of profit maximisation and utility maximisation. Full employment is assumed to prevail both in the city and in the economy as a whole. There are no vacant jobs or any unemployed resources.

3 The Location of the Household

The analysis of the locational choice of the household in this chapter is presented in two ways. In the first part of the chapter we present a diagrammatic account based on the assumption that the consumer attempts to minimise his total location costs, where these are defined as his rent costs plus the direct and indirect costs of travel. This approach has the advantage that it is non-mathematical. Its disadvantage is that, to use it, we have to assume that the householder has only his residential location to decide and that he has already made all his other consumption decisions; in particular, he is assumed to know the amount of housing space he intends to occupy and his valuation of his travelling time.

In the second part of the chapter we use the more conventional economic approach to the theory of consumer behaviour. We assume that the householder attempts to maximise utility, and we determine mathematically the conditions for an optimal location, thus allowing the household to make all his consumption decisions simultaneously. The conditions for optimal location are virtually the same whichever method is used, but the diagrammatic approach is clearer, and the mathematical approach more rigorous.

COST MINIMISATION

It is assumed that the householder attempts to minimise his total location costs and that these location costs are the sum of the total rent that he pays, the direct, financial, cost of travel between his place of work and his residence, and the indirect cost of this journey to work which is the valuation he puts on the time that he must spend on the journey. We make two restrictive assumptions; first, we assume that the number of space units that the householder intends to occupy is known

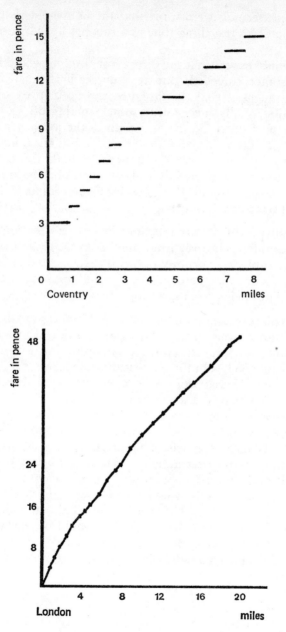

Fig. 3.1   Bus fares in Coventry and London, 1963. *Source:* Wilson (1968).

and fixed; second, we assume that the householder's imputed valuation of his travelling time as a rate per hour is known and fixed.

We would expect that the direct cost of travel would increase with distance travelled, but at a decreasing rate. Empirical evidence suggests that the fare structures set up by transport authorities usually incorporate some tapering off so that the cost per mile decreases as the length of the journey increases. Figure 3.1 shows the cost of travel by bus in London and Coventry as a function of distance. In both cities the fare structures are as expected. This tapering off has been noted in the fare structures for the long-distance transport of goods, where it has been found that:

> typically, tariff per distance unit or zone is steep for the first zone and falls abruptly from the first to the second zone and considerably less abruptly between each succeeding zone (or set of zones) and the one after. Tariff structures are graduated, rates being less than proportional to distance.†

There is no reason to suppose that fare structures relating to the short-distance transport of passengers will not incorporate the same taper as fare structures relating to the long-distance transport of goods. On the empirical evidence we can therefore assume that, viewed from above, the graph of direct travel costs as a function of distance will be convex.

The next step is to derive a graph of the relationship between the indirect cost of travel (the imputed value of travelling time) and the distance travelled. This must be done in stages. To start with we must determine the relationship between distance travelled and time spent in travelling. We would expect, in fact, that travel speeds would increase with the length of the journey, as shown in Table 3.1 from data on journeys in Chicago. Viewed from above, a graph would be concave as in Fig. 3.2, which shows the distance travelled out of London on a radial underground railway as a function of time spent travelling.

Most graphs showing the speed of travel along the radial routes leading out of large cities would have this shape. When travel by only one mode of transport is considered, congestion

† Isard (1956) p. 105. See also Hoover (1948, Chapter 2).

TABLE 3.1

Speed of Travel by Mode by Airline Trip Length, Miles per Hour, Chicago

Airline trip length in miles	Car driver	Suburban railway	Elevated/ subway	Bus
0– 1·9	5·3	3·4	2·8	3·2
2·0– 3·9	8·4	5·8	5·5	5·2
4·0– 5·9	10·6	9·7	7·3	6·6
6·0– 7·9	12·0	10·4	8·6	7·6
8·0– 9·9	13·6	10·9	9·8	8·5
10·0–11·9	14·6	13·1	10·5	10·3
12·0–13·9	15·6	12·7	11·3	11·9
14·0–15·9	17·1	14·2	13·0	11·4
16·0–17·9	18·3	16·4	†	†
18·0–19·9	18·9	15·8	†	†
All trips	11·1	14·4	8·9	6·2

Source: Chicago Area Transportation Study, Final Report, vol. 2, *Data Projections*, July 1960, Table 46.

Notes: 'Speed of travel' denotes airline journey speed, or elapsed time from door to door divided by airline trip length.
'Mode' denotes priority mode. Where two or more modes were used by a traveller, the mode of travel in the linked trip is defined as the mode having highest priority in the following list, taken in the following order: (1) suburban railway, (2) elevated/subway, (3) bus, (4) car driver, (5) car passenger.
† Insufficient data.

near the centre will usually cause short journeys to be under-taken at slower speeds than long journeys. Short rail journeys which may not be slowed by this type of congestion are likely to be slower because stations near the city centre tend to be closer together. Where different modes of transport are used for journeys of different length, the slower form of transport will tend to be used for the shorter journey. Short journeys are undertaken by bus, journeys of medium length by stopping train, and long journeys by express train. The graph would therefore usually have the shape shown in Fig. 3.2, or quadrant A of Fig. 3.3. The speed of travel increases as the length of journey increases. The curve is therefore concave, viewed from above.

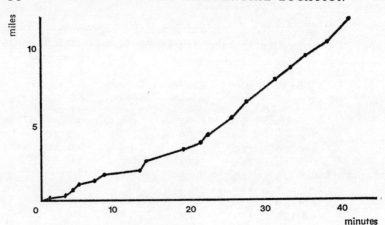

Fig. 3.2 Distance from Charing Cross as a function of time spent in travelling towards High Barnet on the Northern Line. *Source:* London Transport Underground Guide for 1969.

We now assume that the imputed value of travelling time is fixed, as stated above; the graph of the relationship between the indirect cost of travel and the time spent in travelling is then a straight line, as shown in quadrant D of Fig. 3.3.

The graph of the relationship between total indirect costs and distance can be derived from these two graphs. Let the graph in quadrant B of Fig. 3.3 be a 45° line. Given any point on the curve in quadrant D, e.g. point x, we can identify a point (e.g. point x_1) corresponding to it on the curve in quadrant A. At both x and x_1 the same time has been spent in travelling. Therefore, the cost of travel given by x and the distance travelled given by x_1 serve to identify a point x_2 in quadrant C. The dotted line indicates the way in which this can be done geometrically. Point x_2 and other points found in the same way will form a curve with the shape of that in quadrant C, i.e. it will be convex viewed from above.

The graphs of the indirect costs and the direct costs of travel have now both been obtained, and both are convex viewed from above. When added together vertically it is obvious that the graph of the relationship between the total costs of travel and distance travelled will also be convex viewed from above. So that the total location costs can be minimised at a point which is not either the centre or the edge of the city, it is

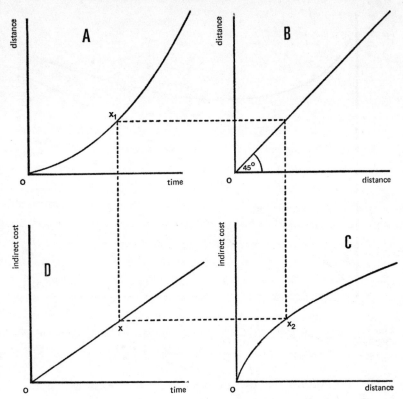

Fig. 3.3   The derivation of the indirect cost of travel as a function of distance.

necessary that the graph of the relationship between location costs and distance should be concave viewed from above and have the shape shown by the topmost curve (*LL*) in Fig. 3.4.†
Since the total travel costs curve (*OT*) has the shape shown by the lower curve in Fig. 3.4, it follows that total rent as a function of distance is given by the area between the two curves. It is obvious that if this were drawn as a graph the curve would be concave viewed from above.

† Although there is no reason why any household should not have its cost-minimising location at the city's centre or its periphery it is obviously impossible for all households to locate at these points. The most general case is therefore an intermediate location.

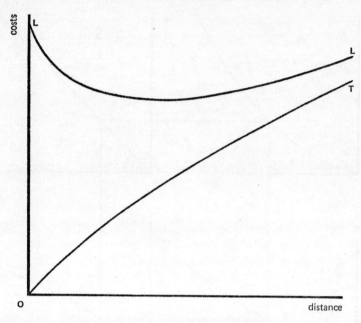

Fig. 3.4 Location costs (*LL*) and transport costs (*OT*) as functions of distance.

Furthermore, since the number of space units which the householder intends to occupy is assumed to be fixed, the total rents payable at each distance from the place of work can be divided by the (fixed) number of space units to obtain a graph of the rent per space unit as a function of distance; this is also concave viewed from above. It follows that for each household to achieve a stable optimal equilibrium location, the rent per space unit must decline at a diminishing rate with distance from the city centre. If it does not do this, if this graph also is convex viewed from above, it would follow that the optimal locations of all the households in the city would either be at the centre or the edge of the city. As a result no household would wish to locate in the intermediate areas of the city, and rents would fall there and increase at the centre and the periphery until the rent gradient (the graph of rent as a function of distance) had ceased to be convex and become concave viewed from above.

It can be seen from Fig. 3.4 that, at the household's cost-minimising location, any small movement closer to or further from the city centre would result in virtually no increase or decrease in its location costs. At points much closer to the centre, however, a movement away from the centre would result in a decrease in location costs because the fall in rents would outweigh the increase in transport costs. At points much further from the centre the increase in transport costs outweighs the fall in rents, and any movement outward would result in an increase in location costs. At the optimal location however the decrease in rents from a move away from the centre is just equal to the increase in transport costs. It is useful to state this result algebraically since it will be needed in later chapters. Let q denote the number of space units occupied by the household, p_k be the rate at which the rent per space unit falls with respect to distance, c_k be the rate at which the direct financial cost of travel increases with respect to distance (i.e. the fare per mile), and $r.v_k$ be the rate at which the indirect cost of travel (the value of travel time) increases with respect to distance, where r is the householder's rate of pay and v_k is a fraction. Then, at the householder's equilibrium location:

$$q.p_k = -(c_k + r.v_k). \qquad (3.1)$$

Thus, the rate of change in total rents with respect to distance is just equal to (the negative of) the rate of change of total travel costs with respect to distance. Alternatively, if p_t, c_t, and $r.v_t$ are the rates of change of rent per space unit, the direct financial cost of travel, and the indirect cost of travel, respectively, all with respect to time instead of distance, the equilibrium location can be restated as:

$$p.q_t = -(c_t + r.v_t). \qquad (3.2)$$

Both these formulations will be used in later chapters.

UTILITY MAXIMISATION

The advantage of the utility-maximisation approach is that the mathematical method allows all independent variables to be handled at once. The householder's valuation of travel time, the number of space units that he occupies, and his location can all be determined simultaneously.†

It is assumed that the householder attempts to maximise a utility function of the form:‡

$$u = u(q, t, w, a_i) \qquad (i = 1, 2, \ldots, n) \qquad (3.3)$$

where q is the number of space units he occupies, t is the time spent travelling from home to work (at the city centre), w denotes the hours of work, and a_i $(i = 1, 2, \ldots, n)$ are all the n other activities of the householder.

The utility function is maximised subject to a budget constraint:

† The two approaches, cost minimisation and utility maximisation, are in a sense equivalent. The relationship between the two methods seems to be similar to that between the primal and the dual in linear programming. It is well known that associated with any linear programming problem there is a dual problem. See, for example, Lancaster (1968, Chapter 3, Section 3.4).

It is apparent that the utility-maximising approach is equivalent here to the primal problem. In a general optimising problem of this kind, however, the Lagrange multipliers are the equivalent of the dual variables in a linear-programming problem, so that the dual is solved and values imputed in the course of solving the primal problem. But one result of the utility-maximising approach is to give an imputed value of travelling time. It should therefore be no surprise to discover that the utility-maximising location is also the cost-minimising location, provided that travel time is given the value imputed to it by the utility-maximising problem. For if the other locations are compared with the optimal location, and the costs of each possible location are compared, it is to be expected that the consumer could not move to a lower cost location, but that the imputed costs of non-optimal locations would be higher. Thus, if all the possible locations are considered as activities, the optimal location will be that for which the imputed costs are lowest, since at any other point the imputed value (in utility terms) will exceed the utility derivable from location there.

‡ It might be preferable to regard this as the utility function of the household, since the house is occupied by the whole household and not only by the person who travels to work. Muth (1969) writes about 'the household', Alonso (1964a) and Wingo (1961a) treat the 'individual' and the 'workers'. Our analysis will follow the latter here, though in later chapters we will usually be writing about the location of the household. There is a distinction between the household and the individuals which comprise it, and this distinction should not be glossed over. On the other hand, an attempt to cope analytically with the distinction between the household and its members might be unsuccessful, could not fail to be complex, and would be unlikely to improve the theory.

$$r.w = q.p(t) + c(t) + \sum r_i.a_i$$

and a time constraint:

$$\bar{T} = w + t + \sum a_i$$

where r is the rate of pay, $p(t)$ the price (per period) per space unit as a function of the time spent in travelling, $c(t)$ the financial cost of travel as a function of the time spent in travelling, $\bar{T}$ the total time in the period, and r_i ($i = 1, 2, \ldots, n$) the cost per hour of each of the n activities a_i.

There are two methods of analysing the way in which the consumer values time in work-type activities, as we have shown elsewhere (Evans, 1972a, 1972c). The first method was illustrated there by an analysis of the valuation of travel time. We assumed that both the origin and destination of the journey were fixed and, hence, that the individual was subject to a third constraint apart from the budget and time constraints.† The second method was used there to analyse the allocation of time to housework. Only two constraints on the individual's behaviour were assumed and it was shown that the housewife would undertake housework only so long as the financial returns from doing so exceeded, or were equal to, her valuation of her own time in doing housework. These financial returns were measured by the amount which she saved by doing housework herself and so neither paying someone else to do it, or buying labour-saving equipment.

In the analysis of the locational choice of the householder we use the second method to determine the imputed value of travel time and, thus, the optimal location. This is necessary because, in this case, the origin and destination of the journey are not both fixed and so the individual is not subject to a third constraint. It is possible because the consumer's valuation of travel time can now be measured by the total rent, net of travel costs, which he saves by travel.

Thus, the first-order conditions for a utility maximum subject to the budget and time constraints are:

$$u_q = \lambda.p(t) \tag{3.4}$$

$$u_t = \mu + \lambda(q.p_t + c_t) \tag{3.5}$$

† This is the only method studied by DeSerpa (1971).

$$u_w = \mu - \lambda \cdot r \qquad\qquad (3.6)$$

$$u_i = \mu + \lambda \cdot r_i \qquad (i = 1, 2, \ldots, n) \qquad (3.7)$$

where $u_q = \partial u / \partial q$, $u_t = \partial u / \partial t$, etc., and μ and λ denote the Lagrange multipliers identified with the marginal utilities of time and money respectively.

These conditions define both the space occupied by the household and its location. The optimal location is the point at which (3.5) is fulfilled. The conditions can also be restated to show that the householder's locational choice implies a valuation of travel time. Division of (3.5) by (3.6) gives, in rearrangement, a new condition:

$$q \cdot p_t + c_t = - \left(\frac{u_t - \mu}{u_w - \mu} \right) r. \qquad (3.8)$$

Let the fraction $(u_t - \mu)/(u_w - \mu)$ be called v_t. Then the total value of travelling time is $v(t) \cdot r$ and the marginal value of travelling time is $r \cdot v_t$. At the optimum, therefore,

$$q \cdot p_t = - (c_t + r \cdot v_t) \qquad (3.9)$$

where v_t is a fraction which may be positive, negative, or equal to zero. Empirical evidence on the valuation of time spent on travelling to work suggests, however, that the value of v_t is usually between 0·20 and 0·30 and that it is reasonably constant.[†] Since r is positive, and the direct financial cost of travel increases with the time spent in travelling so that c_t is positive, it follows that for the household, indeed any household, to be at an optimal location, the term $q \cdot p_t$ must be negative. But only a positive quantity of space units can be occupied so $q > 0$ and $p_t < 0$. The function p is decomposable, for the rent per space unit can be expressed as a function of distance, and distance travelled is a function of time spent in travelling. Therefore $p(t) = p(k(t))$, where $k(t)$ denotes distance as a function of time, and it follows that $p_t = p_k \cdot k_t$.

† See the paper prepared by the Economic Planning Directorate of the Ministry of Transport (1969) in which the results of many British, French, and American studies are abstracted. The fraction may still vary with the method of travel of course. For comment on the effect of this see Chapter 10.

Now it is obviously true that $k_t > 0$, for the distance travelled is an increasing function of the time spent in travelling. Therefore, since $p_t < 0$ it follows that $p_k < 0$. This result can be formally stated as:

Theorem 1. If travel time is valued at a positive fraction of the wage rate, travel costs increase with distance, and distance travelled increases with time spent travelling, then the rent per space unit will necessarily decline with distance from the city centre.

The first-order condition for the optimal location of the household (3.9), then states that the householder chooses his location so that the total rent savings which he would obtain through any small increase in the time spent in travelling between work and home could be just equal to the consequent increase in his total travel costs. The condition can be expressed in terms of distance travelled instead of time spent in travelling. Each of the functions $p(t)$, $c(t)$, and $v(t)$ can be decomposed to read $p(k(t))$, $c(k(t))$, and $v(k(t))$. The new version of (3.9) is then:

$$q.p_k.k_t + c_k.k_t + r.v_k.k_t = 0. \qquad (3.10)$$

If this equation is divided through by k_t, we obtain

$$q.p_k + c_k + r.v_k = 0. \qquad (3.11)$$

This similarly states that, at the optimal location, the reduction in rent costs resulting from a small increase in the distance between residence and workplace would be just equal to the additional travel costs (both direct and indirect). Both these versions of the first-order condition for an optimal location will be used in later chapters to determine the relative locations of different types of households.

The second-order conditions for a utility maximum can be used to show that rents must decline at a diminishing rate with distance from the city centre. We do not need to develop all the second-order conditions. For stability, it is obviously necessary that any large move away from the optimal location should result in an increase in the total rent and travel costs of the householder. If this were not so he would gain by the move, and the equilibrium location would be unstable. Differentiating

(3.10) with respect to time, we obtain as a second-order condition for a stable-equilibrium location:

$$\left(q.p_{kk}.k_t^2 + q.p_k.k_{tt} + \frac{\partial q}{\partial t}.p_k.k_t\right) + (c_{kk}.k_t^2 + c_k.k_{tt})$$
$$+ r(v_{kk}.k_t^2 + v_k.k_{tt}) \geq 0 \qquad (3.12)$$

Our aim is to ascertain the probable sign of p_{kk}. We hypothesise that $p_{kk} > 0$, so that rents decline at a diminishing rate with distance from the city centre. To prove this we must show that the signs of the terms other than $q.p_{kk}.k_t^2$ will be negative or that the terms will be approximately equal to zero. The proof is fairly lengthy.

We start by noting that, if (3.12) is rewritten in the form:

$$k_{tt}(q.p_k + c_k + r.v_k) + k_t^2(q.p_{kk} + c_{kk} + r.v_{kk}) + \frac{\partial q}{\partial t}.p_k.k_t \geq 0$$
$$(3.13)$$

then we know from the first-order condition (3.11) that in equilibrium the expression inside the first set of brackets is equal to zero. We therefore only have to worry about the signs of the other terms in the inequality.

With respect to the final term on the left-hand side of the inequality we have shown above (theorem 1), that $p_k.k_t < 0$, and hence, that any increase in the time spent in travelling would move the householder to a location where the price per space unit is lower. We would therefore expect that there would be a substitution effect with a larger number of space units being bought. Furthermore, since a small move would leave the individual at the same level of real income, it follows that there will be no income effect. If $(\partial q/\partial p)_c$ is the slope of the householder's demand curve for space units at constant real income, then:

$$\frac{\partial q}{\partial t} = \left(\frac{\partial q}{\partial p}\right)_c p_k.k_t.$$

Therefore, the final term in the inequality is equal to:

$$\left(\frac{\partial q}{\partial p}\right)_c p_k^2.k_t^2$$

and its sign is necessarily negative since the demand curve is downward sloping.

The sign of the term c_{kk} must be determined by empirical evidence, and as we showed in the first section of the chapter, the fare structures set up by transport authorities usually incorporate some tapering off so that $c_{kk} < 0$. Indeed, we may note that it would be highly unlikely that $c_{kk} > 0$, for this would mean that short journeys would cost less per mile than long journeys. If this were so it would be possible for passengers to pay for a long journey by dividing it into stages and buying tickets for a series of short journeys. Therefore, it must always be true that $c_{kk} \leq 0$.

This leaves the sign of v_{kk} to be settled, and this is the most difficult to predict. It is best to start with the variation of v with respect to time. We would expect that the householder's imputed value of his time expressed as a rate per hour would either be constant or would increase as the time spent travelling increased. If it is constant then $v_{tt} = 0$ and if it increases $v_{tt} > 0$. Many studies of the valuation of travel time have been carried out, but in only one has it been found that the value of travel time tends to increase with the length of the journey (Lee and Dalvi, 1969; Dalvi and Lee, 1971). I would feel, subjectively, that if the value of travel time does vary, then it varies very little and, hence, that if v_{tt} is positive, then it has a very low value. I would expect that individuals' valuation of their travelling time is fixed in their own minds and would vary little with changes in location. In view of this I shall initially assume that v_{tt} is approximately equal to zero.

If this assumption is made we can discuss the variation in the value of time per mile with an increase in the distance travelled. Since:

$$v_{tt} = v_{kk}.k_t^2 + v_k.k_{tt} = 0 \qquad (3.14)$$

and $k_t^2 > 0$, the sign of v_{kk} hinges on the sign of $v_k.k_{tt}$. Since $v_t > 0$ and $k_t > 0$ and $v_t = v_k.k_t$ it follows that $v_k > 0$. The sign of k_{tt} is also likely to be positive. As we showed earlier in the chapter, the speed of travel on longer journeys is likely to be faster than on short journeys. Indeed, it would be very unlikely that the reverse would be consistently true, for a long journey will not be made more slowly than it would be possible to travel the set of shorter journeys into which it can be broken down. Therefore $k_{tt} \geq 0$. But if $v_k.k_{tt} > 0$ it follows that $v_{kk} < 0$

if (3.14) is to hold. What this means is that, if the value of travel time *per hour* remains constant, the value of travel time *per mile* will tend to fall as the speed of the journey increases, since a smaller period of time will be spent on each additional mile. It should be noted that even if $v_{tt} > 0$ it will still be true that $v_{kk} < 0$, provided that $v_k . k_{tt} > v_{tt}$. In other words, if the increase in speed with additional distance travelled outweighs the increase in the value of time per hour, the value of time per mile still tends to fall with distance.

If the above arguments are accepted it follows that, for the inequality (3.13) to hold and, hence, for equilibrium to be stable, it must be true that $p_{kk} > 0$ since $k_t^2 . q . p_{kk}$ must outweigh the sum of all the other terms which will be negative. This result can be formally stated as:

Theorem 2. If the direct cost of travel increases at a decreasing rate, the speed of travel increases at an increasing rate, and the imputed value of travel time (per hour) does not vary with increases in the time spent travelling, then the rent per space unit will necessarily decline with distance from the city centre at a diminishing rate per mile.

It should be noted that, even if the imputed value of travel time does tend to increase quite strongly with time spent travelling, it may still be outweighed by the three other forces working against it, i.e. the substitution effect, the diminishing cost of travel, and the increasing speed of travel. If, therefore, the value of travel time is not constant the rent per space unit will *probably* decline at a diminishing rate, but will not *necessarily* do so.

As it stands, theorem 2 is not a testable proposition. The space unit is a theoretical construct and so the rent per space unit is also a theoretical construct. One way of letting it be tested would be to lay down co-ordinating definitions allowing the proposition to refer to real houses and flats. This course is followed in Chapter 5 when the proposition is tested in this way. In Chapter 4 a basic theory of the supply of space is presented and this will allow testable propositions to be derived as to the relationship between the value of land and distance from the city centre. Empirical evidence on these propositions is also presented in Chapter 5.

4 The Supply of Space – I

In the previous chapter we derived the conditions for the location of the household to be both optimal and stable, and it was shown that, if certain empirical relationships hold, then these conditions imply that the rent per space unit must decline with distance from the city centre at a diminishing rate. As we showed in Chapter 2 the space unit is a theoretical construct and so this prediction of the theory is not testable in this form. One object of this chapter is therefore to convert the predicted relationship between rent per space unit and distance into a testable prediction about the relationship between land values and distance.

In doing this, we also predict the relationship between density and rent per space unit and hence the relationship between density and land values and distance from the city centre. In Chapter 10 we use the results obtained in this chapter to predict the effects of changes in transport technology.

In this chapter, by making the simplest assumptions, we develop the basic theory of the supply of space. We discuss in Chapter 7 possible modifications to the basic theory of the supply of space.

THE SUPPLY OF SPACE AT A GIVEN LOCATION

The theory of the supply of space developed in this chapter is a variant of the theory of rent in the theory of the firm, and hence, in one form or another dates back to Ricardo.† In the theory as presented here we initially assume the rent per space unit to be given, and derive the supply of space and the value of land from this.

The first step in the analysis is to determine the necessary conditions for equilibrium in the supply of space at a given location. We assume the land at this location (say, one acre)

† See, for example, Henderson and Quandt (1958, pp. 98–101), for a discussion of the theory of rent and the theory of the firm.

to be owned by a ground landlord who rents it to a developer for a fixed ground-rent. When the land has been developed the space is let by the developer to householders. The cost of developing the land varies with the density of development. If we assume that the rent per space unit which householders are willing to pay is fixed, then the only variable within the control of the developer is the density of development. If we assume that the developer attempts to maximise his profits, then we can find the density at which he will do so.

Both plausibility and empirical evidence suggest that a graph of total development costs as a function of density will have the shape shown in Fig. 4.1, i.e. the curve slopes upward, and the slope increases as density increases. We know that at higher densities this is the case. The graph in Fig. 4.1 was drawn using data given by Stone (1959) on the costs of blocks of flats and maisonettes in outer London. The shape of the curve indicates that building cost per space unit (or per standard dwelling)

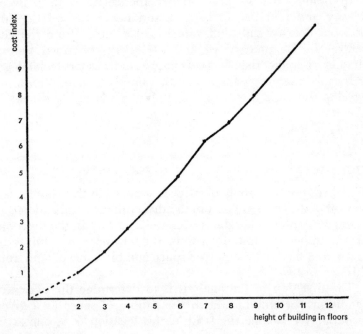

Fig. 4.1 Total building cost as a function of building height. *Source:* Stone (1959).

increases as density increases but at a decreasing rate.† We should expect this to be true for multistorey development. As each additional floor is added to a block of flats the average cost of each flat increases, since the total cost is increased not only by the addition of the topmost floor itself but by the cost of reinforcing the structure to bear the increased weight.

It is less plausible to assume that the graph of cost per space unit as a function of density will have the same shape at lower densities. It is difficult to obtain empirical evidence of this, and the arguments supporting the assumption that it will are not conclusive. It may be that the curve is of irregular shape, sometimes increasing at an increasing rate and sometimes at a decreasing rate. Here, it will be assumed that the graph has the shape indicated in Fig. 4.1 and, hence, that the slope of the curve increases throughout its length.

The developer's profits are calculated by subtracting his total costs from his total revenue from renting the space units. The total costs are the (fixed) ground-rent and the costs of development. In Fig. 4.2, total revenue, total costs, and ground-rent are graphed as functions of density. Since ground-rent is fixed, its graph is shown as a horizontal straight line GG. If the costs of development are expressed as a rate per period, these costs can be added to the ground-rent to obtain the graph of total costs (per period) as a function of density, and this graph is the line GC.

Since it is assumed that the rent per space unit which a householder is willing to pay at that location is fixed, and does not vary with the density of development, then the graph of total revenue as a function of density is a straight line sloping upward from the origin.‡ This graph is OR in Fig. 4.2.

The total profits made by the developer are indicated in Fig. 4.2 by the vertical distance between the total-cost curve, GC, and the total-revenue line, OR. It is obvious that at low densities and at high densities the developer would make losses

---

† The shape of the curve is also confirmed by the data gathered by the Ministry of Housing and Local Government (1966) and given in the Housing Cost Yardstick for public housing schemes at medium and high densities.

‡ In Chapter 7 we consider the implications of an alternative assumption that the rent per space unit does vary with density.

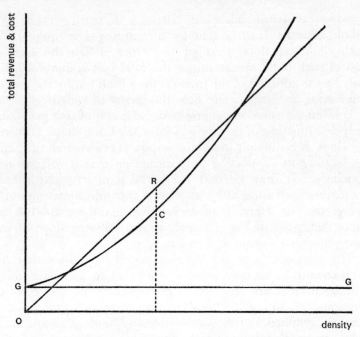

Fig. 4.2   Total cost and total revenue as functions of density.

since the total-cost curve lies above the total-revenue line, i.e. costs exceed revenue. At intermediate densities the developer would make a profit. These profits are maximised at the density at which the vertical distance between OR and GC is maximised (OR lying above GC). This will be true when GC is parallel to OR. At that point any increase in density will lead to an increase in the total costs which will be greater than the increase in the total revenue. Any decrease in density will then lead to a decrease in total revenue which will be greater than the decrease in total costs. At higher densities the marginal cost will be higher than the rent per space unit, while at lower densities the marginal cost will be less than the rent received.

An alternative method of treating the problem diagrammatically is shown in Fig. 4.3. Instead of showing total costs, the diagram shows the average and marginal costs per space unit: the graph of the average ground-rent per space unit as a function of density is shown by the curve $G'G'$. In shape, it is a rectangular hyperbola since the fixed ground-rent is spread

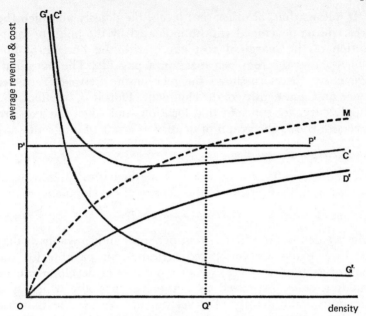

Fig. 4.3   Average revenue and average and marginal cost as functions of density.

over an increasing number of space units as the density increases. The graph of the average cost of development per space unit as a function of density is shown by the curve OD'. The shape of this curve is that indicated by the data collected by Stone (1959) and the Ministry of Housing (1966). The graph of the average total cost per space unit is found by adding $G'G'$ and OD' together vertically, and this is shown by the curve $C'C'$. The average revenue curve is the horizontal straight line $P'P'$ where OP' indicates the rent per space unit which householders are willing to pay at that location.

Curve OM in Fig. 4.3 is the marginal-cost curve, and shows at each density the additional cost of increasing the density by one space unit. As stated above, the density at which profits are maximised is the density at which the marginal cost is equal to the rent per space unit. Therefore, if the rent offered at that location is indicated by OP', the density at which the area will be developed (i.e. the quantity of space which will be supplied per acre) is indicated by OQ'.

It follows that, at other rent levels, the density at which the area will be developed will be indicated by the point of intersection of the marginal cost curve with the horizontal line which shows the rent per space unit payable. The marginal-cost curve therefore shows the relationship between rent per space unit and density of development. Thus it is, in effect, the supply curve for space at that location, and when the axes are reversed, shows the graph of density as a function of rent.

Mathematical Formulation

The same results can be reached mathematically. The total profits, π, are a function of the density of development, d, for:

$$\pi(d) = p \cdot d - g - b(d) \qquad (4.1)$$

where p denotes the (fixed) rent per space unit payable at that location, g denotes the (fixed) ground-rent payable for the acre of ground, and $b(d)$ denotes the cost of development, or building costs (expressed as a rate per period), which is a function of the density of development. We assume that the developer attempts to maximise profits. The first-order condition for a profit maximum is that:

$$\pi'(d) = p - b'(d) = 0$$

or

$$p = b'(d). \qquad (4.2)$$

Equation (4.2) states that at a profit maximum the rent per space unit, p, will be equal to the marginal cost of development, $b'(d)$, which is the conclusion arrived at above. Furthermore, since $b'(d)$ is the first derivative of the function $b(d)$ it follows that (4.2) gives rent per space unit as an explicit function of density and hence gives density, d, as an implicit function of rent per space unit, p. This implicit function gives the supply schedule for space.

The second-order condition for a profit maximum is that:

$$\pi''(d) = -b''(d) < 0$$

or

$$b''(d) > 0. \qquad (4.3)$$

Therefore, so that there should be a profit maximum it is necessary that the second derivative of the building-cost function with respect to density should be greater than zero. This means that the graph of the function must have the shape shown in Fig. 4.1 and derived there from empirical evidence relating to the function.

DENSITY AS A FUNCTION OF DISTANCE

In Chapter 3 it was shown that the shape of the graph of rent per space unit as a function of distance from the city centre could be predicted by the theory. In the first section of this chapter we showed that both the empirical evidence and the conditions for equilibrium in the supply of space dictated the shape of the supply curve for space. In this section these two results are brought together to derive a prediction of the shape of a graph of density as a function of distance.

The first stage is to specify exactly the shape of the graphs of the two functions from which we intend to derive the third. The graph of density as a function of rent per space unit is expected to have the shape shown in quadrant A of Fig. 4.4. This graph, the supply schedule of space, is the marginal-cost curve for the developer. The empirical evidence already cited suggests that the curve will have this shape, in that it will slope upwards but at a diminishing rate.†

The graph of rent per space unit as a function of distance from the city centre is shown in quadrant B of Fig. 4.4. The curve slopes downwards at a diminishing rate, as predicted in Chapter 3. From the rent–distance curve and the rent–density curve we can derive the density–distance curve shown in quadrant D, by means of the 45° line in quadrant C. Thus, if we start with point x on the rent–distance curve in quadrant B, that point indicates a level of rent and its distance from the city centre. Point x_1 in quadrant A indicates the density of development at the same rent level, and so the two together, by giving both a density and a distance, determine the position of the point x_2 in quadrant D. The dotted lines indicate how

---

† Hoch (1969) cites evidence which also suggests that the curve would have this shape. He fits a Cobb–Douglas production function to his data and finds that the coefficient of capital is about 2/3. Fitting Cobb–Douglas functions to Stone's data and the Ministry's data gives coefficients of 0·75 and 0·66 respectively. See next footnote.

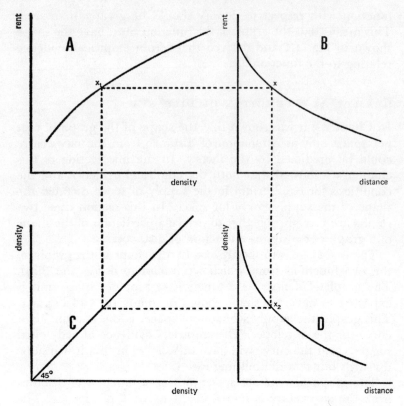

Fig. 4.4   The derivation of density as a function of distance.

this can be done graphically. Other points in quadrant *D* can be found in the same way, and all the points together will form the graph of density as a function of distance. This graph must have the shape shown. Density of space units per acre declines with distance from the city centre, but at a diminishing rate. This result follows from the shape of the two curves from which the graph of density as a function of distance is derived. The graph in quadrant *B* shows that as the centre of the city is neared, rent levels increase at an increasing rate. The graph in quadrant *A* shows that as rent levels get higher, density increases at an increasing rate. Taken together, these two results imply that as the centre of the city is neared, density increases at an increasing rate.

Mathematical Formulation

The mathematical results of the first section of this chapter and of Chapter 3, can be used to predict the shape of the graph of density as a function of distance. Equation (4.2) gives rent per space unit as an explicit function of density. The inverse of this function gives density as a function of rent per space unit. Hence, if we write $p(d)$ for rent as a function of density, and $d(p)$ for density as a function of rent, then:

$$d(p) = p^{-1}(d).$$

The knowledge that $p(d) = b'(d)$ allows us to obtain the first and second derivatives of $d(p)$. Thus:

$$d'(p) = (p^{-1})' \cdot (d)$$
$$= 1/p'(p^{-1}(d))$$
$$= 1/p'(d)$$

so

$$d'(p) = 1/b''(d) \tag{4.4}$$

and

$$d''(p) = -p''(p^{-1}(d))/(p'(p^{-1}(d))^3$$
$$= -p''(d)/(p'(d))^3$$

so

$$d''(p) = -b'''(d)/(b''(d))^3 \tag{4.5}$$

Since we know from (4.3) that $b''(d) > 0$ it follows that $d'(p) > 0$. Further, the empirical evidence cited above suggests that $b'''(d) < 0$ and so it follows that $d''(p) > 0$.†

With these results, we can now show that the density of space units per acre must decline at a decreasing rate with distance from the centre. Let density, d, be stated as a function of distance, k: then density of space units per acre is decomposable into two functions, density being a function of rent per space unit, while rent per space unit is a function of distance, or:

$$d(k) = d(p(k)).$$

† If the production function for space is Cobb–Douglas, the third derivative of the building-cost function will be negative if the exponent of the capital factor in the production function is greater than one-half. As stated in the previous footnote the value of the exponent has been estimated to be between 0·66 and 0·75.

Hence:

$$d'(k) = d'(p(k)) \cdot p'(k)$$
$$= (1/b''(d(k))) \cdot p'(k). \tag{4.6}$$

Since the first term on the right-hand side is positive, as shown in (4.3), and the second is negative, as stated in theorem 1, it follows that:

$$d'(k) < 0 \tag{4.7}$$

and density of space units per acre declines with distance from the city centre.

The second derivative is given by

$$d''(k) = d''(p(k)) \cdot (p'(k))^2 + d'(p(k)) \cdot p''(k).$$

Since, from the analysis above and in Chapter 3, we know that all the terms on the right-hand side of this equation are positive, it follows that density declines at a diminishing rate with distance from centre. Formally, this can be stated as

Theorem 3. If theorem 2 holds, and if the technology of the building industry is such that the average construction cost per space unit increases as the density of space units per acre increases, but at a diminishing rate, then the density of space units per acre will decrease with distance from the city centre, but at a diminishing rate.

LAND VALUES AS A FUNCTION OF DISTANCE

If it is assumed that the ground landlord attempts to maximise the ground-rent which he receives, then we can predict the properties of the functional relationship between land values and distance from the city centre. Suppose that the cost and profit functions at some location are as shown in Fig. 4.2. The profits which would be made by any developer are indicated by the (maximum) vertical distance, RC, between the two curves representing the revenue function, OR, and the cost function, GC, at density Q. The level of these profits depends on the ground-rent obtained by the landlord. If he put the development of the land out to the highest bidder any excess profits would be eliminated, and equilibrium would only be attained when the landlord received a ground-rent which allowed only 'normal' profits to be earned by the developer. The equilibrium position is shown in Fig. 4.5. The total-cost curve is in a position $G'C'$ at which it is tangential to the total-revenue curve, OR, when no 'excess' profits are being earned; this position can be attained by a movement upwards of the line GG in Fig. 4.2 following an increase in the ground-rent. Since 'normal' profits are included in the costs of development, point C on the line GC moves to a position C' which is the same position as point R on the line OR. The optimal density remains the same since it depends on the slopes of the curves being the same, and the slopes are obviously independent of the level of ground-rent.

It can be shown that the level of ground-rents will decrease with distance from the city centre. We start by showing that the

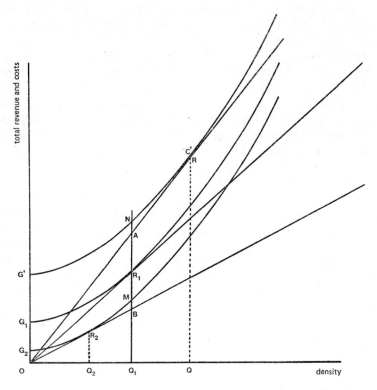

Fig. 4.5   The derivation of land value (ground-rent) as a function of rent per space unit.

level of ground rents declines as the price per space unit declines. Suppose that the rent per space unit is reduced so that the line OR shifts to a new position OR_1 in Fig. 4.5. Losses would then be made by any developer who had to pay a ground-rent of OG', and a new equilibrium would only be attained when the cost curve falls to a new position G_1R_1 tangential to OR_1. The equilibrium ground-rent OG_1 would, of course, be lower than OG' since the curve $G'C'$ has moved downwards. Hence, the level of ground-rents falls as the price per space unit falls. Since the price per space unit falls with distance from the city centre, and since the value of land can be assumed to be the capitalised value of the ground-rents obtainable from the land, it follows that land values will decline with distance from the city centre.

We can go on to show that the level of ground rents will decline at a diminishing rate with distance from the city centre. We assume that rent per space unit falls by the same amount that it was assumed to fall in the preceding paragraph. In Fig. 4.5 this is indicated by the shift of OR_1 to a new position OR_2 where, at any density, the vertical distance between OR_1 and OR_2 is the same as that between OR' and OR_1. Thus, at density Q_1, BR_1 is equal to R_1A. To attain a new equilibrium, the ground-rent must again fall until the total-cost curve attains the position G_2R_2 tangential to the total-revenue curve. The fall in the ground-rent will not be so great as was necessary in the case of the first reduction in the rent per space unit.

To see why this is so, it is necessary to note that the vertical distance between any two of the total-cost curves will be the same at any density, and will be equal to the fall in the ground-rent. Thus, at density Q_1, if the cost curves $G'C'$, G_1R_1, and G_2R_2 cut the vertical line AR_1BQ_1 at points N, R_1, and M, respectively, then $R_1N = G_1G'$ and $R_1M = G_1G_2$. But point N must lie above A, since the total-cost curve $G'C'$ is tangential to the total-revenue curve OAR to the right of A; also point M must lie above B since the cost curve G_2R_2 is tangential to the total-revenue curve OR_2B to the left of B. Hence:

$$G'G_1 = NR_1 > AR_1 = R_1B > R_1M = G_1G_2.$$

Thus, the fall in the level of ground-rents was greater on the first reduction in the rent per space unit than on the second, equal, reduction. Since this result is general, it follows that ground-rents (and therefore land values) fall at a decreasing rate as the price per space unit falls. Since the price per space unit falls at a decreasing rate with distance from the city centre, it follows that land values fall at a decreasing rate with distance from the city centre.

Mathematical Formulation

The above result can be obtained mathematically. In (4.1) above, the relationship between the developer's profits and density of development was stated to be:

$$\pi = p.d - g - b(d). \tag{4.8}$$

If it is assumed that the ground landlord is able to extract all excess profits in ground-rent, then $\pi = 0$. If we still assume that the developer attempts to maximise profits (i.e. attempts to prevent losses), then we know from (4.2) that he will develop to the density such that $p = b'(d)$. Substituting in (4.8) and rearranging, we obtain an equation giving ground-rents, g, as a function of density, d:

$$g(d) = b'(d).d - b(d).$$

The first derivative of the function g is then:

$$g'(d) = b''(d).d + b'(d) - b'(d)$$

or

$$g'(d) = d.b''(d). \tag{4.9}$$

But density, d, is a function of distance, k, and we want to obtain information on g as a function of k. Thus, from (4.9) and (4.6) above:

$$g(k) = g(d(k))$$

and

$$g'(k) = g'(d(k)).d'(k)$$

$$= d.b''(d).(1/b''(d)).p'(k)$$

Therefore:

$$g'(k) = d(k).p'(k) \tag{4.10}$$

and, since $p'(k)$ is negative (from theorem 1) and $d(k)$ is positive, it follows that:

$$g'(k) < 0$$

and ground-rents (and land values) decline with distance from the city centre. Differentiating (4.10) with respect to k we obtain the second derivative:

$$g''(k) = d(k).p''(k) + d'(k).p'(k).$$

Since both $d(k)$ and $p''(k)$ are positive and $d'(k)$ and $p'(k)$ are both negative (from theorems 1 and 2 and equation (4.7)) it follows that:

$$g''(k) > 0$$

and ground-rents (and land values) decline at a diminishing rate with distance from the city centre. It is noteworthy that this result is not dependent on the technology of the building industry. In this, it differs from the result of the analysis of the shape of the density gradient in the preceding section. The shape of the land-value gradient is therefore less dependent on empirical properties than the shape of the density gradient. The result can be stated formally as:

Theorem 4. If theorem 2 holds, then the value of land will decline at a diminishing rate with distance from the city centre.

CONCLUSIONS

The result stated in theorem 4 is highly satisfactory from the viewpoint of testing the theory. This result is not stated in terms of theoretical constructs but in terms of land values and distance on which data can be obtained. This data can thus be used to confirm or refute the predictions of the theory, and in Chapter 5 we shall review the studies that have been made to show that the predictions of the theory are confirmed.

Certain aspects of the supply of space have been glossed over in this chapter. Thus, we have not dealt with the problems which may arise because buildings typically have long lives, and both the rent per space unit and the developer's profits will vary over the life of the building. These problems will be dealt with in Chapter 7 where we will also consider the effects of different types of land and property tenure; also, the possibility that households may not be indifferent to the density of development but may be willing to pay differing prices for development at differing densities.

5 Empirical Evidence on Land Values, Property Values, and Population Density

In the two preceding chapters we set out a theory of household location and a theory of the supply of space. It was shown that these theories, if correct, implied that the value of land, rent per space unit, and density per space unit should all decline at a diminishing rate with distance from the city centre. In this chapter these implications of the theory will be tested.

In the first section it is shown that the published empirical evidence on the relationship between land values and distance confirms the theoretical prediction. In the second section the predicted relationship between rent per space unit and distance is converted into a testable relationship between property values and distance, and this is tested against data obtained on property values in a sector of London. The theoretical prediction is not refuted by this test. In the third section it is shown that the predicted relationship between density of space units per acre and distance can be used to predict the probable relationship between population density and distance, and that this probable relationship is that shown by the considerable body of empirical evidence.

THE LAND-VALUE GRADIENT

The conclusion reached at the end of the previous chapter is clear and uncompromising. If the theory presented in Chapters 3 and 4 is correct, and if the necessary empirical relationships exist, then the value of land will decrease as distance from the city centre increases, but the rate of decrease will diminish as distance increases. The exact relationship in a particular case will depend on the characteristics of the city and its inhabitants, i.e. transport costs, transport speeds, building costs, building technology, consumer tastes.†

Obviously the value of land in any city is not a function of distance from the city centre alone: there are other exogenous variables, e.g. distance from suburban centres. If the assumptions underlying the theory are correct, however, distance from the city centre or central business district (CBD) will be one of the most important explanatory variables in any statistical analysis of land values in the whole city. If this is found to be true, it would tend to confirm the assumption that distance from the CBD must be one of the most important factors in any explanation of intra-urban patterns of residential location.

Empirical evidence

Studies of land values in many cities provide a considerable body of evidence to support the theory. The results of several studies at various dates have been brought together by Colin Clark (1966, 1967), and the graphs of the log of land value as a function of distance from the city centre are shown in Fig. 5.1. It can be seen that the function is negative exponential for Perth and Vancouver, while for the other cities the curvature is more pronounced. Only the data for Los Angeles suggest that in that city there may be no decline in land values with distance from the CBD. Statistical studies of the value of land in other cities also support the theory. In Okayama (Hawley, 1955) and Seattle (Seyfried, 1963) the functional relationship was found to be double logarithmic; in Lyons and Marseilles (Granelle, 1968) a negative exponential function was fitted to the data. The empirical results from all the cities, with the exception of Los Angeles, confirm the predictions of the theory. Furthermore, it can be seen that distance from the CBD must be one of the most important explanatory factors in all the cities for which the relationship between land values and distance is graphed in Fig. 5.1.

Even in the case of Los Angeles, statistical analysis suggests that distance from the city centre is an important explanatory

---

† It is not predicted that the graph of the function will take a particular form; nor is it predicted that the function will be negative exponential or double logarithmic. Clark (1967, p. 382n) notes that it has been suggested the function should be double logarithmic.

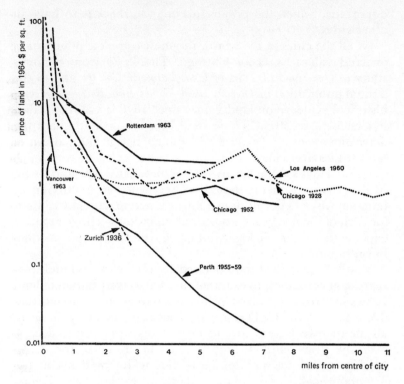

Fig. 5.1 Land values in some major cities. *Source:* Clark (1966).

factor. In a recent study Eugene Brigham (1965) shows that 'though the *plots* [of land value against distance] do not reveal clear land value gradients, *these gradients are nonetheless present in the Los Angeles market.* The plots are two-dimensional and do not include the very important amenity factor [i.e. the condition and status of the neighbourhood]. When amenities are brought in, as they are in the regression equations, then a very strong negative relation is seen to exist between value and distance from the CBD' (p. 333, Brigham's italics). In this study Brigham assumed a linear relationship and did not test for curvature. Nevertheless, the results show that, in Los Angeles, distance from the city centre is an important factor in determining the value of land, even though, in a notoriously dispersed city, it has not the overriding importance which the graphs in Fig. 5.1. show it to have in other cities.†

Of all the cities in the world, the most complete information on land values exists for Chicago. Thanks to Homer Hoyt's study of *One Hundred Years of Land Values in Chicago* and to the annual publication of *Olcott's Land Values Blue Book of Chicago*, historical evidence on land values since 1836 is available. Two independent researchers have used this data to study both the determinants of land values and the change in the pattern of land values over time.

Using Hoyt's data for five dates between 1836 and 1928, Mills (1969) explicitly tested to see whether a curvilinear function was a better fit than a linear function. He showed that, for each of the five years, non-linear equations – either negative exponential or double logarithmic – gave much better fits than linear equations.

Yeates (1965) used data derived from *Olcotts'*, also multiple-regression equations, to examine the spatial distribution of land values in 1910 and every tenth year up to 1960. He found that distance from the CBD was a significant explanatory factor in all the regressions, but that its importance appeared to decline over the years until, in 1960, both the distance from Lake Michigan and the percentage of non-white residents in the neighbourhood, were more important explanatory factors. However, the percentage of the variance explained by all the variables included in the regressions fell over time, from 77 per cent in 1910 to 18 per cent in 1960. This may be due to the fact that Yeates' study area was only the City of Chicago: in 1910 the area of the City may have been approximately the same as the metropolitan area, but by 1960 this would certainly not be true. Since we would expect the slope of the land-value gradient to become less steep over time because of improvements in transport technology, it follows that, over time, a smaller proportion of the decline in land values due to distance from the CBD would lie within the City of Chicago. Thus, if the regression equations for later years were fitted to data which included

† The partial correlation coefficient between land value and distance to the CBD lies between –0·49 and –0·89 in the regressions carried out by Brigham. This suggests that, even in Los Angeles, distance to the CBD may be the most important factor determining the value of land.

the land values in the newer suburbs of the City, it is probable that distance from the CBD would again be the most important variable.†

In a study of land values in Britain, P. A. Stone (1964, 1965) found 'for London and Birmingham [a negative] exponential relationship between price per acre and distance from the centre of the region' (1965, p. 5). The relationship held up to about 30 miles from the centre of Birmingham and up to about 60 miles from the centre of London. The observations used by Stone as data were the auction prices of residential-building sites with outline planning permission for erection of a given number of dwellings; both the distance from the city centre and the permitted density in dwellings per acre appear as independent variables in Stone's regression equations. But it is questionable whether density should be treated as an independent variable, as by Stone, or whether it should be treated as a dependent variable determined simultaneously with land values, as the theory implies. If the British planning authorities exert stringent planning control and set residential densities which are independent of the densities which would be set by market forces, then Stone would be correct. On the other hand, if either planning authorities are unable to exert control, or set densities which are more or less the same as those which would be set by the market, then Stone is incorrect.

The evidence on this would suggest that, in fact, planned residential densities are highly correlated with existing densities, which it is plausible to assume are largely unplanned. From a sample of 107 British planning authorities Lever (1971) shows that the residential density in new development was highly correlated with existing residential density, more than 50 per cent of the variation in the new (planned) densities being 'explained' by variation in the existing densities. Furthermore, Stone himself notes that 'densities are highest in the centre of cities and decline systematically as distance from the centre

† Note that Yeates' statistical analysis shows that land values fall with distance from the CBD, even in 1960. Thus, it is not true to claim, as Clark (1967) does, that the 'recent study by Yeates . . . shows how the situation has almost inverted itself since 1910'. The coefficients of the log of distance in the regressions for 1910 and 1960 are negative, and significantly different from zero at the ·05 level.

increases' (1965, p. 5). Thus it is doubtful whether Stone is correct in treating density as an exogenous variable. On the other hand, it is obvious that, whether or not he is methodologically correct, his findings confirm the theoretical predictions, for in both London and Birmingham land values decline with distance from the CBD but at a diminishing rate.

THE PROPERTY-VALUE GRADIENT

In Chapter 3 it was shown that if certain empirical relationships hold, the rent per space unit will decline with distance from the city centre but that the rate of decline will diminish as distance increases (theorems 1 and 2). This prediction can be turned into a testable prediction by making use of the first of the postulates by which we gave a partial interpretation of the space unit in Chapter 2, i.e. 'if a house or flat or other dwelling is identical to a house or flat or other dwelling at another location, then it has exactly the same number of space units.' Thus, if we specify a type of dwelling as carefully as possible and find the value of this type of dwelling in different parts of the city, this will give a property-value gradient. If we assume the number of space units in the type of dwelling to be constant, then, because the property-value gradient has a certain shape, the gradient of rent per space unit will also have this shape, since we could turn property values into a capital value per space unit by dividing by the number of space units in this type of dwelling. The capital value can, of course, then be turned into an annual rent by dividing by r, where the rate of interest is $100r$ per cent.

For instance, Frieden (1961) obtained data on the price of three types of apartments in New York City, namely two-bedroom, one-bedroom, and one-room apartments. The price of each of the three types as a function of distance from the CBD is shown in Fig. 5.2. If we can assume either that, for example, each one-bedroom apartment consists of approximately the same number of space units as any other one-bedroom apartment, or that variations in the number of space units are not correlated with distance from the city centre, then Frieden's results tend to confirm the predictions of the theory.

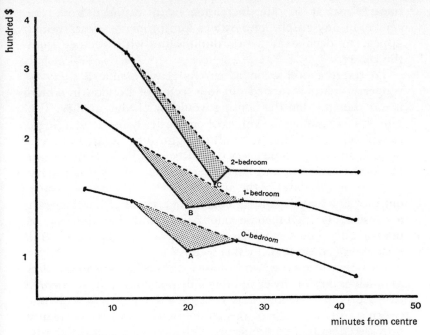

Fig. 5.2 Monthly rents and travel times from centre of New York region, including poor environments in 15 to 25-minute zone, 1958–1959. *Source:* Frieden (1961). City of New York Department of City Planning rental inventory of buildings completed 1958–1959; 14,152 dwelling units in 136 apartment developments in 'good environments' plus 2,500 units in 15 developments in 'poor environments'. Median rents and median times of poor environment sample plotted at *A, B,* and *C.*

The price of property falls with distance from the centre but at a diminishing rate.

On the other hand, Frieden suggests that in New York the values of properties located within the inner suburbs are depressed by a poor environment. This suggests that the theory must be tested more stringently than before, since it could be argued that the curvature in the land-value and property-value gradients may not be the result of the economic forces postulated in the theory but could be due to the presence of decayed inner suburbs in each city. The presence of decaying areas near to any property would, it is argued, depress the value of that property; hence, although the 'real' land value or property-value gradient may be such that values decline linearly or at a rate increasing with distance from the centre, in any graph, land values and property values would appear to decline at a rate diminishing with distance from the centre.

To test the prediction as carefully as possible a study of property values is carried out in a sector of London in which it was thought that the housing would be of high quality. The aim was to examine the values of properties lying along a single radial route and which were all thought to be situated in areas of good standing; these areas were Regent's Park, Hampstead, Highgate, Finchley, and Barnet. Some subjectivity is necessary in assessing the status of the areas, but there is evidence to bear out the view that these areas constitute a sector of London with a consistently high-income population, and that the ring of poor-quality housing which otherwise surrounds the central area is broken through by this sector.†

Variation in the size and quality of housing was minimised as much as possible by examining only one house type – namely, two-bedroom leasehold flats constructed after 1960 with between 92 and 99 years of the lease to run – and estate agents provided details of 29 properties of this type for sale or recently sold in the sector in September 1967. The price of each property was adjusted by adding to it the capitalised value of the ground rent and, where necessary, by subtracting from it the estimated value of garages and cloakrooms. The adjusted asking price in pounds, P, was then regressed against distance in miles, D from the city centre, the total area of the three main rooms in square feet, A, and the number of expired years on the lease, T. The equation with the highest level of explanation was the form:

$$\log P = 1 \cdot 2018 + 0 \cdot 2808 \log A - 0 \cdot 6534 \log D - 0 \cdot 0179 T$$

$$R^2 = 0 \cdot 84 \qquad (5.1)$$

or

$$P = 15,920 . A^{0.28} . D^{-0.65} . 0 \cdot 96^T \qquad (5.2)$$

† See, for example, Map 5–7, 'Household Income by Traffic District' in *The London Traffic Survey*, vol. 1 (London County Council, 1964) and the map of housing conditions appended at the end of *A Report on Greater London and Five Other Conurbations: Census 1951* (General Register Office, London, 1956).

The coefficient of A is not very reliable, and variation in A accounts for very little of the variance:† variations in the prices of properties are almost entirely accounted for by variations in distance, by the expired period of the lease, and by other factors which have not been taken into account. The partial R^2 for the three independent variables are 0·09 for log A, 0·74 for log D, and 0·31 for T. Thus, the equation is in the form predicted by the theory, namely, the predicted price of the properties falls with distance from the city centre but at a diminishing rate, and this is so even in the absence of any decayed property at intermediate distances. In order to see whether, in the actual relationship between property prices and distance, there was any curvature which was not explained by the equation fitted to the data the residuals from the regression were calculated. It was found that the predicted prices were generally below the actual prices of properties close to the centre and at the edge of the built-up area, while the predicted prices of properties at intermediate distances were generally above the actual prices. This indicated that the amount of curvature in the relationship was even greater than could be explained by the log linear function fitted to the data. An analysis of variance showed this residual curvature to be statistically significant.‡ These results serve to confirm the theory. Moreover, distance from the city centre explains almost three-quarters of the variation in the value of the type of property studied, and this again confirms the assumption that distance from the centre is an important determinant of patterns of residential location.§

One point arising from the empirical results is worth noting. The coefficient of T in (5.2) is 0·96, and this indicates that the value of the property falls each year by 4 per cent of its value at the beginning of the year. This agrees with the result of an

---

† This is largely due to the fact that, in standardising the number of rooms in the flat, we also limit the variation in the area of the flat. When the number of rooms is not standardised, area becomes a much more important explanatory variable. See, for example, the studies of London house prices by Wabe (1971) and Lane (1970).

‡ $F = 4·04; F_{0.05} = 3·42.$

§ But, of course, it should be noted that the influence of differences in the size and quality of housing or in its environment has been largely eliminated.

American study quoted by Grebler, Blank, and Winnick (1956, Appendix E) which showed that the value of a property appeared to decline by about 4 per cent per annum in the first few years of its life. But the dependent variable in the regression equation is the asking price, and it may be considered possible, even probable, that properties being sold by owner-occupiers would be advertised at a price some £300 above the expected selling price or market price. The high estimated rate of depreciation suggests that this is not the case. Indeed, it would seem to imply that owner-occupiers may underestimate the true value of their properties.

There are two possible reasons for this high rate of depreciation. First, since developers know the state of the market better than owner-occupiers do, the latter may underestimate the rate of inflation and fail to realise the true market price of their properties. The seller may be content to make a paper profit on the sale and may not attempt to maximise this profit. Second, and more probably, the rate of depreciation of the value of the property may be very high in the early years of the life of the property, higher than the average rate of depreciation of the property over its whole life; the reasons why this should be so are discussed towards the end of Chapter 7.

THE DENSITY GRADIENT

The third theoretical prediction considered in this chapter is that, provided certain plausible empirical relationships hold, the density of space units per acre will decrease with distance from the city centre but at a diminishing rate. As it stands, this prediction is untestable since we are unable to measure density in terms of space units, the latter being theoretical constructs.† Nevertheless, we can turn it into a prediction that population density will decline with distance from the CBD and that the rate of decline will probably decrease as distance increases. The first part of this prediction, at least, will then be testable.

† It is worth noting, however, that Mills (1969) found that density in terms of floor space per acre of developed land declines with distance from the centre of Chicago. Furthermore a log linear equation gave a better fit than an exponential or linear equation.

Let $D(k)$ denote density in persons per acre, then $D = d/q$, or

$$D(k) = d(k)/q(p(k))$$

where d denotes density in space units per acre, k denotes distance from the centre, q denotes space units per person, and p denotes rent per space unit. Then:

$$D_k = \frac{1}{q^2}(q \cdot d_k - d \cdot q_p \cdot p_k) \qquad (5.3)$$

where $D_k = D'(k)$, etc. Of the terms within parentheses, $q \cdot d_k < 0$, since $d_k < 0$ (theorem 3), and $d \cdot q_p \cdot p_k > 0$, since $p_k < 0$ (theorem 1), and $q_p < 0$ (i.e. the demand curve is downward sloping). Hence, $D_k < 0$, and population density must decline with distance from the city centre. This theoretical prediction is unequivocal and obviously testable.

Differentiating (5.3), we obtain:

$$D_{kk} = \frac{1}{q^2}(q \cdot d_{kk} - d \cdot q_{pp} \cdot p_k^2 - d \cdot q_p \cdot p_{kk}) - \frac{2}{q^3}(q \cdot d_k - d \cdot q_p \cdot p_k).$$
$$(5.4)$$

The second expression in parentheses in (5.4.) was shown in the preceding paragraph to be negative. Of the three terms in the first set of parentheses, $q \cdot d_{kk} > 0$, since $d_{kk} > 0$ (theorem 3), $d \cdot q_p \cdot p_{kk} < 0$, since $p_{kk} > 0$ (theorem 2), and $q_p < 0$ (the demand curve is downward sloping), but the sign of the term $d \cdot q_{pp} \cdot p_k^2$ is indeterminate, since the sign of q_{pp} is indeterminate. If $q_{pp} < 0$, then $d \cdot q_{pp} \cdot p_k^2 < 0$, and it follows that $D_{kk} > 0$. We can conclude that $D_{kk} > 0$, unless $q_{pp} > 0$ *and* the positive effect of all the other terms does not outweigh the negative effect of $-d \cdot q_{pp} \cdot p_k^2$. We can conclude that it is highly probable, but not certain, that the rate of decline of population density will decrease as distance from the city centre decreases.

There is a wealth of evidence to show that population density does decline with distance from the centre and that it does, in fact, decline at a diminishing rate. This abundance of evidence would seem to be due to the availability of Census data on population densities, to Colin Clark's (1951) finding that the relationship between urban population densities and distance from the city centre appeared to be negative exponential, and to his claim that this appeared 'to be true for all times and all places studied, from 1801 to the present day, and from Los Angeles to Budapest'.

In the sixties this finding was tested exhaustively. Berry, Simmons, and Tennant (1963), on the basis of their own studies and those of others, claimed that 'almost a hundred cases are now available, with examples drawn from most parts of the world for the past 150 years, and no evidence has yet been found to counter Clark's assertion of the universal applicability of [the negative exponential equation]. To be sure, the goodness of fit of the model varies from place to place, but in every place so far studied a statistically significant negative exponential relationship between density and distance appears to exist' (p. 391).

If the theory is correct this is what one would expect. The negative exponential function has the property that it slopes downwards at a diminishing rate. Therefore it should fit the data for any city quite well. There is no reason to suppose however that it should fit the data for all cities exactly. So Muth (1969), when he fitted negative exponential functions to the data for 46 major U.S. cities, found that 'the linear regression [of log density on] distance alone explains slightly less than half of the variation in population density among census tracts' and that 'too many significant deviations from linearity were observed to attribute to sampling variability' (p. 145).

Other researchers have gone on to fit other types of function with similar characteristics to the negative exponential function. For example, Newling (1969) suggested that the density function should be a quadratic exponential,

$$D(k) = ae^{bk-ck^2}$$

instead of negative exponential,

$$D(k) = ae^{-bk}$$

where $D(k)$ is population density at distance k from the city centre, and a, b, and c are constants.

Finally, Casetti (1969) fitted various types of functions (potential, square-root negative exponential, negative exponential, and negative exponentials of second and third degrees) to several sets of data and concluded that the investigation has shown that the strikingly dominant feature of urban population distribution the decline of densities with distance from city centres, and that negative exponential functions of degree from 1 to 3 are better suited to represent the density decline in central urban areas, while potential functions are best suited to outskirts and rural fringes. However, all the functions investigated give excellent fits, so that good results obtained with one particular family of functions is, in itself, not a good enough reason for preference. Perhaps the negative exponential function of first degree, that lies somewhere between exponential functions of higher degree, better suited to central areas, and functions of lower degrees, suited to peripheral areas, deserves its popularity because it is a compromise solution (p. 111).

These empirical results entirely support the theory and in no way refute its predictions. Densities do decline with distance from the centre, and the relationship between the rate of decline and distance which theoretical deduction showed to be highly probable does, in fact, appear to hold.

CONCLUSIONS

In this chapter we have tested some predictions of the theory of residential location and shown that these predictions are not refuted by the available evidence. Moreover, the successful results of these tests demonstrates that the assumptions of the theory appear to be reasonable. If it were not reasonable to assume utility maximisation by consumers, profit maximisation by landlords and developers, and a single major workplace in a city, then the predictions derived from the theory might be refuted.

The evidence set out in this chapter therefore allows us to have some confidence in the plausibility of the theory. In later chapters we shall develop the theory, and derive further explanations of observed patterns and further testable predictions. The next chapter is wholly theoretical, however, and is devoted to the development of tools of analysis to be used in later chapters.

6 The Bid-Price Curve and Market Equilibrium

The empirical studies in the preceding chapter were, in a sense, a digression from the main thread of the argument; however, they demonstrated the validity of the theories in Chapters 3 and 4 of the demand for, and the supply of, space. In this chapter we bring together these theories to show how demand and supply are equated when the market is in equilibrium. Since space is supplied at every location in the city there is no single market price. We cannot therefore take over, unchanged, the analysis of market equilibrium which is customary in standard microeconomic theory where goods are assumed either to have no locational dimension or to be brought to, and sold at, a market at a particular location. In the case of habitable space in the city the market has to be thought of as a number of interconnected submarkets, each at a different distance from the city centre and each having a different equilibrium price and a different equilibrium quantity traded.

In this chapter we make no distinction between long-run and short-run equilibrium. We assume that long-run equilibrium is instantaneously achieved. Of course, this is an unrealistic assumption to make about the market for space; buildings, once built, are highly durable, and small fluctuations in the price of space will not cause old buildings to be demolished and new buildings to be constructed, whether at a higher or a lower density. On the other hand, the assumption is not critical to the analysis in this chapter and most of the later analysis, and it considerably simplifies the argument. In a later chapter on the supply of space some problems raised by the durability of buildings will be discussed.

This chapter is in three sections. In the first section we develop our tool of analysis, the bid-price curve, in the second, we show how two households locate relative to each other, and in the third we show how market equilibrium is attained. The chapter is short because both market equilibrium and the bid-price curve have been discussed in depth by Alonso (1964a, pp. 59–100), and there is little to add to his analysis. It is essential, however, to include at least a brief outline of the subject, both to draw together the threads of the arguments followed in the preceding chapters, and to provide a basis for the investigation of patterns of residential location in succeeding chapters.

THE BID-PRICE CURVE

The bid-price curve was originally defined and developed by William Alonso (1964a).† He uses the concept in the analysis of industrial and agricultural location but we are only concerned here with its use in the theory of residential location.

A bid-price curve for a particular household may be obtained in this way. We assume that the tastes, income, family characteristics, etc. of the household are given. Suppose this household is initially only allowed to locate at the centre of the city and that the rent per space unit at this location is given. The household maximises its utility at this location subject to usual time and budget constraints on its behaviour. Hence, associated with the price of space to the household at that location is a particular utility level. Obviously, a higher price would be associated with a lower utility level (i.e. it would be preferred less), and a lower price would be associated with a higher utility level (i.e. it would be preferred more). Suppose, now that we choose another location within the city, but that this time, the household is asked to state the price per space unit at that location, at which it would be indifferent between that location and the city centre. Whereas, in the first instance, the rent level was fixed and the utility level found, at the suburban location the rent level is found and the utility level fixed equal to the maximum attainable at the centre. It will be remembered that one of the assumptions of Chapter 2 – which has not yet been relaxed – is that the city is located on the usual, flat homogeneous plain; hence, any difference between the

† The basis of the idea is present in earlier studies in the theory of agricultural location. For example, Isard (1956, p. 195) uses a 'rent function', and this is similar to Hoover's use of a curve of 'ceiling rents'(1948, p. 95). See also Lösch (1954, pp. 36–67).

price per space unit at the two locations is solely due to differences in their location relative to the workplace at the city centre. Therefore, because of travel costs – both direct and indirect – the price per space unit at the non-central location must be lower than the price at the centre if the household is indifferent between them.

This procedure could be carried out for locations at any distance from the city centre. At each location, the household could be asked to state a price per space unit at which it would be indifferent between that location at that price, and location at the centre at the given price. Thus we can obtain a schedule of prices and locations or distances from the city centre, and call this a bid-price schedule. If enough information is obtained and the bid-price schedule is detailed enough we can obtain a function stating bid prices as a continuous function of distance from the city centre, the utility level of the household being held constant. The graph of this function would show the bid price decreasing with distance from the city centre. Alonso calls this line the bid-price curve. It is obvious that the bid-price curve is similar in derivation and interpretation to the indifference curve used in the analysis of the foundations of the theory of consumer behaviour.†

Just as there is, for each household, a complete set of indifference curves covering all possible combinations of com-

---

† Some economists, e.g. Mishan (1961), have suggested that the study of the economics of consumer behaviour should leave aside the analysis of the foundations of the theory and start with the assumption of a downward-sloping demand curve. But, in the study of the theory of residential location, the demand curve is a concept which is of little use. Other tools of analysis have to be developed – hence, the concept of the bid-price curve.

Now it might be possible to start with the assumption of a downward-sloping bid-price curve but, as we shall show, the work which we carried out in Chapter 3 using the foundations of consumer-behaviour theory allows us to state the exact slope of this curve. Furthermore, the analytical foundations of the economics of consumer behaviour are necessary to tie together bid-price curves and demand curves into one coherent, general theory. This means that either concept can be used with more confidence than would otherwise be the case, since the user is reassured by the knowledge that both can be derived from a more general theory and, hence, that his explanation of a particular facet of consumer behaviour is part of a more general explanation.

modities so there is, for each household, a complete set of bid-price curves covering all possible pairs of prices and distances. Thus, by setting some higher price per space unit at the city centre, the household could be induced to reveal another bid-price curve. Since this curve would be higher than the first bid-price curve derived it would be associated with a lower utility level. If the price per space unit set at the centre were lower than the first price given, the bid-price curve obtained would be associated with a higher utility level. For the same reason that indifference curves cannot intersect, so bid-price curves cannot intersect.† For any household, we can draw in a

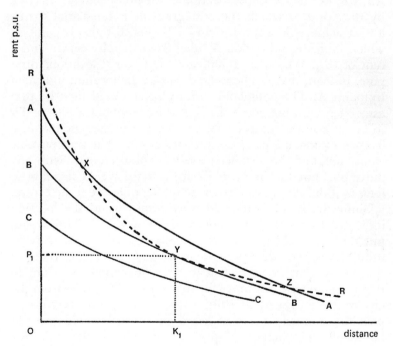

Fig. 6.1   The optimal location of the household.

† Thus, suppose that bid-price curves AC and BC intersect at C, and points A and B indicate the bid prices on each bid-price curve at some location. The householder is indifferent between A and C, and C and B and, hence, between A and B. But this is impossible because A and B indicate different rents at the same location. Therefore, it is impossible for bid-price curves to intersect.

diagram a few of the household's set of bid-price curves. In Fig. 6.1 we draw three such curves, AA, BB, CC, each indicating by its decreasing height above the horizontal axis that it denotes an increasingly preferred rent level and, hence, an increasingly greater utility level.

The equilibrium location of the household

If both the rent gradient and the household's set of bid-price curves were known, then the optimal location of the household could be shown diagrammatically, as in Fig. 6.1. There the rent gradient RR is shown, together with three bid-price curves AA, BB, CC. The optimal location of the household is indicated by the point at which the rent gradient is tangential to the lowest achievable bid-price curve. In Fig. 6.1 this is point Y, which indicates a location K miles from the city centre and a rent of OP_1. Why does Y indicate the optimal location? Suppose, instead, that the household was at the location indicated by point X. The household's utility level would then be that associated with the curve AA. But the household would prefer to be on a lower bid-price curve, and it can exercise this preference by moving out from the city centre. The rent payable would fall, and the distance travelled would increase with any move out, but the household's utility level would increase so long as it did not move further than OK_1 miles from the centre.

Similarly, if the household were originally at the location indicated by point Z any short move toward the centre would put the household on a lower bid-price curve and increase its utility level, even though it would be paying a higher rent per space unit. This would be true for any move inwards so long as the household did not move further in than OK_1 miles from the city centre. From K_1 no shift on to a lower bid-price curve is possible, and any move outward or inward would put the household on a higher bid-price curve. Thus, Y indicates the household's equilibrium rent per space unit and location, and this location is optimal because the rent gradient is tangential to the lowest achievable bid-price curve at Y.

The slope of the bid-price curve

We can describe some of the necessary characteristics of a set of bid-price curves. In Chapter 3 the conditions for the household's location to be optimal were determined mathematically. It was shown (3.11) that the first-order conditions for an optimal location could be stated as a requirement that

$$q.p_k + c_k + r.v_k = 0 \qquad (6.1)$$

where q denotes the number of space units occupied by the household, p_k the rate at which the rent per space unit declines with distance from the city centre (i.e. the slope of the rent gradient), c_k the rate at which the direct financial cost of travel increases with distance travelled, r the rate of pay of the householder, and v_k is a fraction such that $r.v_k$ indicates the householder's valuation of his travel time (per mile) as a fraction of his rate of pay.

For our present purposes, (6.1) is more usefully written as:

$$p_k = -\frac{c_k + r.v_k}{q}. \qquad (6.2)$$

This equation gives the first-order condition for an optimal location in terms of the slope of the rent gradient. It states that, at the optimal location, the slope of the rent gradient must be equal to the negative of the rate of change of total travelling costs divided by the number of space units occupied by the household at that location.

In this form, the equation can be used to describe the slope of the bid-price curve, for in the preceding section it was demonstrated that the rent gradient must be tangential to a bid-price curve at the optimal location. It follows, therefore, that at the optimal location the slope of the rent gradient, p_k is, equal to the slope of the lowest achievable bid-price curve, p_k^*, or

$$p_k^* = -\frac{c_k + r.v_k}{q}. \qquad (6.3)$$

In later chapters this result will be used to derive testable predictions about the patterns of residential location in large cities. Here we will use it only to describe some of the characteristics of a set of bid-price curves.

Equation (6.3) appears to give the slope of the lowest bid price curve achieved at the optimal location. In fact, the result is more general. We do not know which is the lowest achievable bid-price curve and which is the optimal location until we know the rent gradient. But since any rent gradient is possible,† it follows that the rent gradient could be tangential to any point on any bid-price curve. Hence the slope at any point of any bid-price curve must be equal to $-(c_k + r.v_k)/q$, though the values of c_k, v_k, and q will, of course, vary systematically from point to point and curve to curve.

Using this formula, we can demonstrate that the slope of any particular bid-price curve must become less steep as distance increases. First, as we showed in Chapter 3, c_k declines as distance increases. Second, if it can be assumed that the householder's valuation of his travel time is a constant rate per hour, and if the speed of travel increases with distance, his valuation of time will be a decreasing rate per mile, since a shorter period of time is spent in travelling each additional mile. Again, this was demonstrated in Chapter 3. Thirdly, since the price per space unit falls as distance increases, and no income effect is possible since the householder moves along a single bid-price curve and, hence, remains on the same indifference curve, it follows that q increases as distance increases. Only a substitution effect is possible and q cannot decrease as distance increases.

Thus, the two terms in the numerator of the fraction, c_k and $r.v_k$, decline as distance increases, while the denominator, q, increases. Therefore, the absolute value of the fraction, and thus the slope of any bid-price curve, must diminish as distance from the city centre increases. Bid-price curves must therefore be drawn with a diminishing slope, as in Fig. 6.1.

The formula can also be used to show the way in which the slope of a higher bid-price curve will differ from that of a lower curve. At any given distance from the city centre, if one bid-price curve is lower than another, then by definition, the price of space is lower. It is plausible to assume that space is not an inferior good and so the lower bid-price curve will be associated with the purchase of a greater number of space units. Hence, the denominator q of the fraction $(c_k + r.v_k)/q$ will be greater. On the other hand, c_k will be constant, and it is plausible to assume that the householder's marginal valuation

† With the qualification that it must slope downwards at a diminishing rate for the optimal location to be also an equilibrium location.

of time (per mile) will also be constant.† Thus the numerator of the fraction is constant. It follows that the lower the bid-price curve at any location, the lower will be the value of the function $(c_k + r.v_k)/q$ and, hence, the less steep will be the slope of the curve. From this it follows that a characteristic set of bid-price curves should be as shown in Fig. 6.1. The vertical separation between any two curves must decrease as distance from the centre of the city increases.

THE LOCATION OF TWO HOUSEHOLDS RELATIVE TO EACH OTHER

So far, in this chapter, we have only described the set of bid-price curves of a single household. In this section we consider the sets of curves of more than one household, and show that of any two households the one with the steeper bid-price curves will locate nearer the centre. This follows from the fact that at any household's optimal location the rent gradient must be tangential to one of that household's bid-price curves. Thus, in Fig. 6.2. the rent gradient RR is drawn as a tangent to the bid-price curve AA at point X, where X indicates a location OK miles from the city centre and a rent OP per space unit for a particular household.

Two other bid-price curves, BB and CC, are also drawn in Fig. 6.2; both pass through point X but come from the sets of curves of two further households.‡ Curve BB is steeper than curve AA, and it follows that X does not indicate the optimal location of this second household since its bid-price, BB is not tangential to RR at X, It can be seen immediately that the second household must find its optimal location nearer the city centre, since any move to a new location nearer the centre would move it on to a lower bid-price curve. Thus, the household with the steepest bid-price curve BB will locate nearest the centre. Similarly, X does not indicate the optimal location

---

† These two statements are correct only if we are considering a given city with a given transport system. Since the transport system which exists in a city is not independent of its size, the sets of bid-price curves of a household will vary from city to city.

‡ For simplicity it is assumed that X is the only point of intersection of the three bid-price curves AA, BB, and CC.

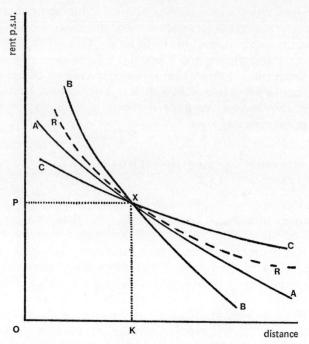

Fig. 6.2   The location of different households.

of the third household, since its bid-price curve CC is not tangential to RR at X; it can be seen that the third household will find its optimal location further out, since any move outward will move it on to a lower bid-price curve. The household with the least-steep bid-price curve CC will locate furthest from the centre. These results are completely general since the same analysis can be carried out for any location and for all households' bid-price curves. Thus the households with the steepest bid-price curves will locate nearest the centre. This result will be used to derive empirically testable predictions in later chapters.

MARKET EQUILIBRIUM

So far we have always taken the rent gradient as given, and then determined the optimal location of a household or households. The bid-price curve is a tool of analysis which allows us to move somewhat away from this partial-equilibrium approach towards a general-equilibrium approach. Using it we can show how the rent gradient is determined by the competition for space when not only one household, but every household, in the city is free to vary its location.

The most elementary case

Suppose that all the households in the city have sets of bid-price curves which appear identical when drawn. It should be noted that this assumption does not mean that a given bid-price curve represents the same utility level for each household. No interpersonal comparisons are implied. It is assumed only that, if each household is asked to outline the bid-price curve associated with a given rent level at the city centre (or some other location), each household's bid-price curve will be identical.

Suppose that Fig. 6.3 shows some of the set of curves. It can be easily seen that the rent gradient of the city in market equilibrium must lie wholly along one of the set of curves (though

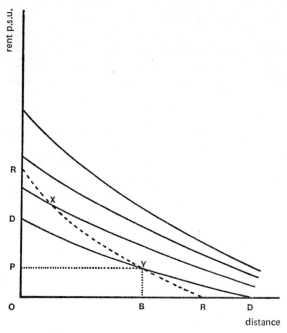

Fig. 6.3 The rent gradient in disequilibrium.

not necessarily one of those shown). If it did not, and it was continuous, there must be points at which the gradient is not tangential to any bid-price curve, as RR is not a tangent at X and Y. Hence, no one would choose the locations associated with these points, and the gradient cannot be an equilibrium rent gradient. Furthermore – and this applies even if the gradient is discontinuous – if it did not lie wholly along one bid-price curve some householders would not be on the lowest achievable bid-price curve and would gain by relocation. For example, in Fig. 6.3 let the equilibrium rent gradient be DD and the actual rent gradient be RR. Those located less than OB miles from the centre could gain by movement outward. Indeed, as Fig. 6.3 stands, all households except those at the extreme edge of the city would gain by moving to the extreme edge (OR miles from the centre) since this would be the rent–distance combination lying on the lowest bid-price curve. There would thus be a fall in demand in the inner areas and an increase in demand at the edge. Similarly, there would thus be a fall in rents in the inner areas of the city, and increases in rents at the periphery, coupled with new development between R and D miles from the centre. This process of adjustment would go on until the rent gradient lay wholly on a bid-price curve.

In the example above we stated that line DD was the equilibrium bid-price curve. Of course, any of the set of bid-price curves could be the equilibrium curve. Which one it will be will depend on the size of the population of the city and their demand for space.

The higher the rent gradient, the greater the area of the city and the greater the density of space units at each location; hence, a higher rent gradient is associated with a greater area of space for occupation of households, and the larger the population of the city, the higher is the rent gradient.

The analysis can be stated mathematically. The density of space units, d, at any location depends on the price per space unit, p, ruling at that location, or

$$d = d(p).$$

If a possible rent gradient is stated as

$$p_A = p_A(k) \tag{6.4}$$

where k denotes distance from the city centre, it follows that the density at distance k_1 is given by

$$d_A(k_1) \equiv d(p_A(k_1)).$$

The total number of space units at that distance from the centre will be

$$2\pi k_1 . d_A(k_1)$$

so that the total number of space units available for occupation in the city will be

$$2\pi \int_0^K k . d_A(k) \mathrm{d}k$$

where K is the radius of the city given by the distance from the centre at which $p_A(k)$ equals the rent ruling on the plain surrounding the city. For (6.4) to be the equilibrium-rent gradient the quantity of space units supplied must be equal (at all distances from the centre) to the number demanded. We will assume that the demand curve for space is the same for each consumer and does not vary with his location.

$$q = q(p) \tag{6.5}$$

where q, the number of space units occupied by each household, is a function of rent per space unit, p. From the rent gradient (6.4) and the demand function (6.5) another function:

$$q_A = q_A(k)$$

can be obtained giving the number of space units demanded per household as a function of distance. At a point at distance k_1 from the centre the density of households to the acre will be equal to the density of space units divided by the number of space units occupied by each household, or

$$\frac{d_A(k_1)}{q_A(k_1)}.$$

The total number of households at that distance from the centre will be:

$$\frac{2\pi k_1 . d_A(k_1)}{q_A(k_1)}$$

and the total number of households who can live in the city will be:

$$N_A = 2\pi \int_0^K \frac{k.d_A(k)}{q_A(k)} dk.$$

If N_A is equal to the total population of the city then $p_A = p_A(k)$ is the equilibrium-rent gradient. If N_A is less than the total population of households wishing to live in the city, the total number of space units supplied at each location will be too low. Demand will exceed supply and the rents at each location will be bid up until demand equals supply and the rent gradient lies along a higher bid-price curve. If N_A is greater than the total population of households, the supply of space units will exceed demand at each location, and the rent gradient will fall until it reaches equilibrium along a lower bid-price curve.

The more complex case

If we drop the simplifying assumption that all the households in the city have identical sets of bid-price curves, then, as we have shown, land nearest the centre of the city will be occupied by those with the steepest bid-price curves. The rent gradient will not lie along a single bid-price curve, but will be made up of sections of the lowest attainable bid-price curves of all the households in the city. As in the simplest case, the height of the rent gradient (i.e. the rent level at any given location) will depend on the size of the population to be housed, but now it will also depend on the proportion of the population of each type and the area of land required to house them.

For example, suppose for simplicity that there are only two different types of bid-price curve; a possible rent gradient would be the curve ABC in Fig. 6.4 where AB is a segment of one of the set of steeper bid-price curves and BC is one of the set of less-steep curves. ABC is the equilibrium rent gradient only if one group of households has its demand for space met within the circle of radius OR miles and if the other group can be accommodated in the ring of width CR miles and inner radius OR miles.

Even if only one section of the market is in disequilibrium the whole rent gradient has to shift upwards or downwards. Thus, if the demand for space by those with the steepest bid-

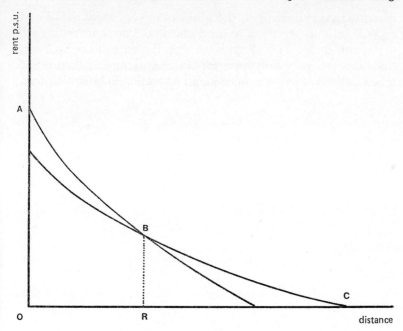

Fig. 6.4 The equilibrium rent gradient – two types of household.

price curves exceeded the supply within the circle of radius *OR*, the price of space would tend to increase in that area, and the segment *AB* of the rent gradient would shift upwards (and outwards). The effect of this would be to increase the area occupied by those with the steepest bid-price curves at the expense of the other households. Therefore, the supply of space in the outer ring would be reduced and demand would exceed supply. The segment *BC* of the rent gradient would have to shift upwards (and outwards) until a new market equilibrium was reached.

SUMMARY AND CONCLUSIONS

In this chapter we have presented a brief account of the derivation and properties of the bid-price curve as a tool of analysis and have used it to show how equilibrium is attained in the market for space. The discussion has no special claim to originality, and the subject has not been investigated in depth.

Alonso (1964a) deals with the problem in very great detail and there is little to be added to his discussion. The refinements are dealt with there. This chapter presents the core of the analysis, since this is a necessary basis for the empirical and theoretical investigations of patterns of residential location in later chapters.

7 The Supply of Space – II

The basic theory of the supply of space was set out in Chapter 4. From it we derived three testable predictions about the relationship between land values, property values, density, and distance from the city centre, and these predictions were confirmed by the evidence presented in Chapter 5. From this we concluded that the theory is reasonably realistic, and in Chapter 6 the theory was used as the basis for a discussion of the characteristics of equilibrium in the market for space in the city. If the market for space is considered in more detail, however, certain of the assumptions made in Chapter 4 may be thought unrealistic. In this chapter we digress from the main argument to consider the implications of altering two of them, in order to determine the possible implications for the theory of residential location.

In the first place, it has so far been assumed that households at a given location are willing to pay the same rent per space unit whatever the density of development. This assumption may not be correct. For example, households may prefer low densities to high and may therefore be willing to pay higher rents for space provided at lower densities. The implications of varying this assumption are considered in the first section of the chapter.

Secondly, all the variables so far used have been expressed in terms of their value per unit period of time. Not only rent per space unit and ground-rent but development costs also, were expressed in this way. In effect the durability of housing was completely ignored; the theory was stated as if housing were both produced and consumed within a single time period. This form of the theory may be adequate in most cases, but some features of the housing market will necessarily be ignored. In the second and third sections of the chapter the basic theory is presented using capital values instead of values per period. This presentation allows us also to discuss two further problems in the theory of the supply of space, the effect of leaseholds and the existence and nature of the filtering-down process.

VARIATIONS IN RENT WITH DENSITY

Just as the theory of the supply of space set out in Chapter 4
corresponds to the theory of the firm in perfect competition, so
it will be seen that the theory of the supply of space set out here
corresponds, in many respects, to the theory of the mono-
politically competitive firm.

Suppose, initially, that the rent per space unit which householders are willing to pay at a given location does *not* vary
with density. Then the equilibrium situation will be as in
Fig. 7.1. The graph of the average cost of development per
space unit as a function of density is shown by the curve *OD*.
Curve *OM* is the marginal-cost curve and shows, at each
density, the additional cost of increasing the density by one
space unit. The rent offered at that location is *OP* and is
invariant with density, as indicated by the horizontal line *PP*₁.
Profits will be maximised when price equals marginal cost; on
the diagram this is indicated by the intersection at *L* of the

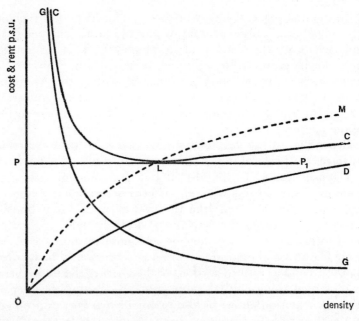

Fig. 7.1 Equilibrium in the market for space (price per space unit
constant).

marginal-cost curve OM and the (horizontal) demand curve PP_1. The average ground-rent per space unit is shown by the rectangular hyperbola GG, and the average total costs per space unit by the curve CC (found by adding average development costs to average ground-rent). Since the landowner sells the land to the highest bidder, no developer earns 'excess' profits, and the curve CC, in equilibrium, has PP_1 as a tangent; the landowner obtains the average ground rent denoted by GG with the result that only development at the profit-maximising density yields a profit, and development at any other density yields a loss.

Now suppose that, in fact, householders are not indifferent to the density of development of an area but dislike high density development. Furthermore, the higher the density of development in an area, the lower the rent per space unit they are willing to pay in that area. We may think of the area in question as a housing estate of several hundred houses or flats, and the density of development as the number of dwellings to the acre on the estate; we assume that housing densities outside the estate do not affect the rents that the occupiers of dwellings on the estate are willing to pay.

A firm developing the estate as a whole faces a downward-sloping average-revenue curve in choosing the optimal density for the estate. The equilibrium solution is shown in Fig. 7.2. Associated with the average-revenue curve, PP_1, is a marginal-revenue curve, PR, indicating at each density the additional revenue obtained by the developer from an increase in density of one space unit per acre. The cost conditions represented by the average-development-cost curve, OD, and the marginal-cost curve, OM, are identical to those represented in Fig. 7.1. It can be easily seen that the developer will maximise his profits by development at the density at which marginal revenue equals marginal cost, the density indicated by the intersection of the marginal-cost and marginal-revenue curves in Fig. 7.2. If the landowner maximises his ground rents, the developer will again earn no excess profits so that in equilibrium the curve CC must be tangential to PP_1. The point of tangency L therefore indicates both the rent and the density of development of the estate when it is developed by a single firm.

Thus far it can be seen that the theory is in many respects

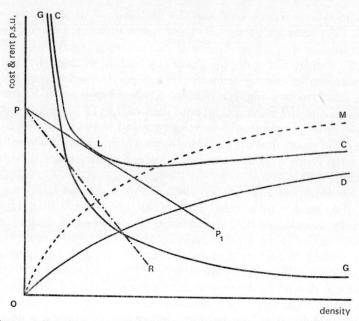

Fig. 7.2   Equilibrium in the market for space (price per space unit dependent on density).

identical to the theory of the monopolistically competitive firm. However, the two theories differ in one important respect. The average-revenue curve used here is not identical to the demand curve of monopolistic competition but is entirely different. In this analysis we assume that perfect competition exists at each density in that the household has open to it the possibility of occupying space in other developments at all possible densities. The slope of the average-revenue curve indicates the fall in price which is necessary to compensate the household for the higher density of development in the estate. Thus, the slope of the average-revenue curve in this case does not arise from any 'monopoly' but to compensate the residents of the estate for the social cost of the increased density of development.† It follows from this that the equilibrium solution shown

† The problem is similar to that of road congestion, and as such is an application of the theory of economic clubs. See Buchanan (1965). A mathematical analysis of the problem in relation to variation in densities within the city has been carried out by Mirrlees (1972).

in Fig. 7.2 is Pareto optimal, whereas if we were considering a monopolistically competitive firm the profit-maximising output would not be Pareto optimal.

The density will be Pareto optimal only in certain circumstances, however. The density which determines the rent per space unit that a household is willing to pay is, as stated above, the density of development of the whole estate, and not just the density of development of the single site within the estate on which the household happens to live. If, as we assumed above, the density of development of the sites surrounding the estate can be ignored, and if the estate is developed by a single developer then the development density will be Pareto optimal. But if the site is developed, or is redeveloped, piecemeal by different developers, then the resulting density may not be Pareto optimal.

Suppose, for example, that a single site is to be redeveloped within the estate. When the estate was originally developed the original developer faced the situation shown in Fig. 7.2 and maximised his profits in the way shown there. Suppose now that a single site in the area is later to be redeveloped and that cost and demand conditions have remained exactly the same. So far as the developer of the site is concerned, if the site is small enough the density of the estate will remain virtually unchanged whatever the density at which he redevelops the *site*. Hence, so far as he is concerned he faces an almost horizontal average-revenue curve through L, shown as $P'LL'$ in Fig. 7.3. This average line $P'LL'$ must be distinguished from the average-revenue line PLP_1. The line PLP_1 shows the average revenue of a developer as a function of the density of development of the *estate* if he is developing the whole estate; the line $P'LL'$ shows the average revenue of a developer as a function of the density of development of a *site*, assuming that the estate is already developed at the density P_1 indicated by point L and that the developer is redeveloping a single small site within the estate.

The horizontal axis in Fig. 7.3 therefore indicates both density of development of the estate (for PLP_1) and the density of redevelopment of a site (for $P'LL'$). The redeveloper of a single site maximises his profits by setting marginal revenue equal to marginal cost, and so his profit-maximising density

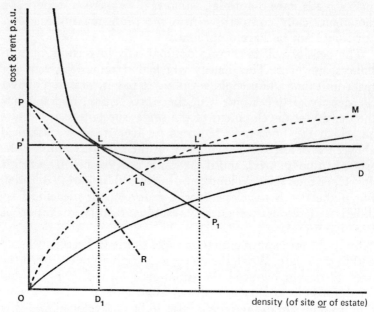

Fig. 7.3   Piecemeal development and the market for space.

is indicated by point L' at the intersection of his average-revenue line and the marginal-cost curve.

Obviously, redevelopment raises the average density of the estate. With more and more piecemeal redevelopment the rent per space unit obtainable at the site will fall as the average density increases. Stable equilibrium can be reached only when the rent and density for all developments on all sites are those indicated by point L_n.† Both the average-revenue curve for area, PLP_1, and a horizontal average-revenue curve for a site

† I first thought that L_n would be reached through average rents falling continuously and densities on the estate increasing continuously along the path indicated by the line LL_n as redevelopment continued. John Harris (M.I.T.) has convinced me that this is not so. The density of redevelopment will fall as indicated by the line $L'L_n$ as redevelopment continues. When L_n is reached it seems quite possible that the estate will be composed of original low-density development and new high-density development. If the original sites are redeveloped at a higher density this will raise the average density, even though they are redeveloped at the average density. It can be seen that oscillation in the average rent and density of the estate can occur about the equilibrium point.

cut the marginal-cost curve at L_n; hence, any redevelopment of a site would take place at the same density as the average density of development of the estate. L_n would also indicate the equilibrium rent and density if the estate were originally developed piecemeal by separate developers. Each developer of a site would maximise his profits without taking into account the effects of his own decision on the profits of the other developers. It is to be noted that the ground-rent of the estate when it is developed piecemeal will be lower than when it is developed by one developer. If the site were owned by one landlord he could therefore ensure that the site was developed as a whole by setting the price of the land (or ground-rent) high enough.

The analysis is, in this respect precisely analogous to the analysis of the theory of the firm. In the single-developer case (= monopolistic competition) the land values (or profits) are higher than in the many-developer case (= perfect competition). On the other hand, the welfare implications are the complete reverse. Development by a single firm results in a Pareto optimal density, while piecemeal development leads to a non-Pareto optimal result. This occurs because the average-revenue curve that we have assumed to face the single developer is not the same as the demand curve facing the monopolistic firm. The slope of the average-revenue curve reflects the social cost of increased density. It indicates that any increase in density would result in the residents of an estate being made worse off unless they were compensated by a reduction in the rent per space unit.

The marginal-revenue curve therefore represents the revenue resulting from the rent of one additional unit *less* the social cost of the increase in density of that one unit as reflected in the fall in the rent per space unit which could be charged. In Figs. 7.2 and 7.3 the vertical distance between the marginal-revenue curve, PR and the average-revenue curve, PLP_1, indicates the marginal social cost of any increase in density. Since the marginal-cost curve, OM, denotes the marginal *private* development cost of any increase in density, it follows that the rent per space unit is equal to the sum of the marginal social cost and marginal private cost only at the density indicated by L. At L_n, rent per space unit equals marginal private cost but marginal social cost is ignored. Hence, the density which would result from piecemeal development is not Pareto optimal, while the density which would result from development by a single developer is Pareto optimal.

Note that this analysis provides a justification for the practice of laying down the maximum residential densities at which local authorities will allow areas to be developed (or redeveloped). In the case illustrated in Fig. 7.2 for example, the authority would set the density indicated by point L as the maximum residential density at which development would be allowed. Development of the whole area by a single developer would not be affected but piecemeal development would be forced to conform to the same standard.

The above analysis is purely theoretical. Though it is plausible to assume that there is some downward slope to the average-revenue curve facing the developer of an area, there is no direct evidence to suggest that the slope is anything but negligible. The analysis does generate one testable prediction, however, namely that small-scale developments are likely to be at higher densities than large-scale developments. A study by E. A. Craven (1969) of private residential development in Kent between 1956 and 1964 provides some evidence bearing on this proposition. Craven concluded that

> There is a strong correlation between the size of development and the type of housing being built. The large development tends to consist of semi-detached and terrace houses (or semi-detached bungalows); the small development of detached dwellings or high density flats and maisonettes. This is especially true of semi-detached houses; 76 per cent of those appearing in the sample were found in developments of over 36 dwellings. Detached dwellings which were built in large developments tended quite often to form a minor part of an estate dominated by semi-detached or terrace houses. This sprinkling of detached houses is included presumably to increase sales through raising the potential social status of the estate (p. 8).

To some extent Craven's conclusions support the analysis above. To some extent they do not. Small developments consist both of the highest densities (flats and maisonettes) and of the lowest densities (detached houses and bungalows), whereas large developments are carried out at medium densities. Craven's evidence certainly does not disprove the hypothesis but neither does it confirm it.

It is certainly true that local authorities lay down maximum densities for the development of areas but it is doubtful whether the reasons presented here were the reasons for the imposition of maximum densities, and, if they were, whether the local authorities or their town planners had any evidence, other than intuition, to justify their imposition. Jane Jacobs (1961, Chapter 11), for example, has argued that planners have confused high density due to a large number of persons per room with high density due to a large number of dwellings per acre, and have attempted to cure the overcrowding implicit in a high density in terms of persons per room by restrictions on the density of dwellings per acre.

The importance of the factors treated in the analysis presented in this section cannot therefore be confirmed. Certainly, if it were true that the rent per space unit that people were willing to pay at any given location was greatly affected by the density of development in the surrounding area, then the theory of residential location presented in the earlier chapters might need to be altered, but so far as can be seen this alteration would not greatly affect the results obtained so far and to be obtained in later chapters. On the other hand, it should be remembered that, as stated in Chapter 2, the systematic variation in density with distance from the city centre may mean that the economic welfare of those living in the inner city is reduced relative to those living in the suburbs, so that the residential patterns predicted in later chapters may not be (normatively) optimal.

An earlier, alternative, version of the method of analysis used above is used by G. M. Neutze (1968) to explain the construction of tall apartment blocks in the suburbs of American cities. He argues that 'since land is not expensive, it appears to be the demand rather than the cost side that mainly explains the appearance of high-rise apartment buildings on the urban fringe', and he suggests that many residents are willing to pay 'a height premium' to live on the upper floors of tall blocks. Furthermore, 'because the roof can be developed for a garden and swimming pool, the height premium, or part of it, can be obtained from all tenants' (p. 84). If a height premium exists, and this premium increases with the height of the building, it follows that the developer faces what appears to be an upward-sloping demand curve. The equilibrium situation is shown in Fig. 7.4. The marginal-revenue curve PR now lies wholly above its associated 'demand curve', PP. Again the profit-maximising density and rent are indicated by point L at the tangency of the average-total-cost curve and the average-revenue curve. This point lies immediately *below* the point of intersection of the marginal-revenue and marginal-cost curves.

The apparent contradiction between Neutze's analysis and the one presented earlier may not be real. Neutze is concerned with the determination of the height of a single building; we are concerned with the determination of the optimal density of an area. It would be quite possible for a developer to decide that his profits would be maximised if a whole site were developed at a low density with a single high building on it. Thus, the two versions of the analysis are perfectly consistent.

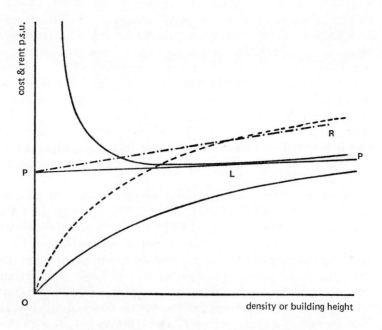

Fig. 7.4 Equilibrium when there is a height premium.

As Neutze remarks:

> building height is not necessarily closely related to density. In fact, in outer suburban areas this is unlikely, if for no other reason than that high buildings require a great deal of parking, and this can be most cheaply provided in open car parks. There is no premium for garages with a view, though tenants may be willing to pay something to have their cars stored inside. But very frequently the attraction of the location is partly that the apartment building has a considerable open area around it for outdoor living and recreation (p. 85).

CAPITAL VALUES − I

So far, housing costs and housing prices have been expressed in terms of their value per unit period of time. But to ignore the durability of housing is to ignore an important feature of the housing market and to close our eyes to any consequences which it might have. Of course, it would be possible to wish away much of the problem by assuming that any building has an indefinite life and that both rent per space unit and maintenance do not depend on the age of the building. Then both can be capitalised by dividing by the rate of interest. Thus the capitalised value of a rent per space unit, p, would be p/i where the rate of interest is $100i$ per cent per year. In certain circumstances this may be a useful assumption to simplify the problem,† but in most cases these assumptions are not only false; they might lead to incorrect conclusions. Both the costs of maintenance and the rent per space unit may vary over the life of the building, and the building may therefore have a definite life.

In the remainder of this chapter we therefore investigate the implications of assuming that the rent of a building and the cost of maintenance vary over time. One type of model is used

† For example, Needleman (1969) makes the assumption in a study of 'the comparative economics of improvement and new building'. He assumes that 'improved' buildings will have a limited life but that new building will have an infinite life. This assumption is obviously very favourable to new building, but this does not matter in the context since, despite making this assumption, Needleman successfully demonstrates that it may still be cheaper to provide housing by improvement rather than rebuilding.

in this section as the basis for the study of the effect of lease-holds, and a more mathematically sophisticated one appears in the next section where we discuss filtering down and depreciation.

Suppose, to take a simple case, that the variation in rent and maintenance over the life of a building is as shown in Fig. 7.5. For clarity we assume that a developer rents the site from a landlord for a ground-rent, constructs the building on the site, and lets it to householders, the rent being adjusted each year to allow for changes in the condition of the property. Curve *II* in Fig. 7.5 is the graph of the net income per annum from the property as a function of the age of the property. The curve *IR* is a graph of the gross rental income p.a. obtained from the property, also as a function of the age of the property. The vertical distance between *II* and *IR* shows the annual expenditure on the maintenance of the building by the developer.

It seems plausible to assume that the rent declines over time and that the expenditure on maintenance increases. Line *GG* shows the ground-rent which has to be paid to the landowner each year. The ground-rent is assumed not to vary over time. The developer will demolish the building and rebuild when his net income from the property fails to cover the ground-rent. In Fig. 7.5 this is indicated by the net-income curve, *II*, cutting the line *GG* from above when the building is aged *T'* years. Of course, if the developer owned the land himself then the line *GG* would indicate the opportunity cost of the land to him. If he had not developed the site himself, he could have leased the land to another developer at the rent *OG*.

Note that demolition costs must be taken into account in determining the optimal life of the building. This can be clearly seen if we assume that after *T'* years, when the ground-rent equals or exceeds the net income from the property, the developer relinquishes his interest in the property and allows the building as well as the land to revert to the ground landlord. The landlord has to demolish the building and clear the site before he can lease the site again to a new developer at the same ground-rent, *OG*. Hence, if he demolishes the building immediately, his net income from the new lease would be equal to *OG* less the interest on the cost of demolition. It is therefore profitable for the landlord to postpone demolition until the annual net income from the property becomes equal to the ground-rent *less* the interest on the cost of demolition (i.e. until, in Fig. 7.5, the line *II* cuts a horizontal line lower than *GG*).

When the building is planned, not only must its life be anticipated but its density must be determined. The determination of the optimal density presents no new problems. The developer's profit will be maximised at the density at which the marginal-capital cost per space unit is equal to the capitalised revenue derived from the sale or rent of a space unit. The graphs used would be the same as those used in Chapter 4, and the course of the argument would be the same.

Mathematical Formulation

The rent per space unit, p, is assumed to be a function of its age, t, or $p = p(t)$. The present value of the gross revenue from one space unit from the start of development $(t = 0)$ to age T is equal to:

$$\int_0^T p(t) . e^{-it} dt$$

where the rate of interest is equal to $100i$ per cent per annum. The present value of G_T, the price per acre at which the site

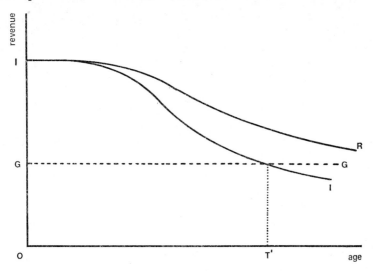

Fig. 7.5   The optimal life of the building.

can be sold in T years' time is equal to $G_T . \mathrm{e}^{-iT}$, and the present value of D_T, the costs of demolition in T years' time, is equal to $D_T . \mathrm{e}^{-iT}$.

The total cost of development is equal to the sum of the initial cost of the acre of land to be built on, G_0, the cost of construction, $B(d)$, as a function of the density of development and the present value of the costs of maintenance. The latter are assumed to be a function both of time and the density of development. To simplify the problem we assume that the form of the function is such that the costs of maintenance per annum for the development are equal to $f(d) . m(t)$ where $f(d)$ is a factor varying with density, and $m(t)$ is a factor varying with age. The present value of the costs of maintenance over the life of the building are therefore equal to:

$$f(d) \int_0^T m(t) . \mathrm{e}^{-it} \mathrm{d}t.$$

We assume that the developer sets out to maximise the present value of his profits, π, over the life of the asset, and

$$\pi = d \int_0^T p(t) . \mathrm{e}^{-it} \mathrm{d}t - f(d) \int_0^T m(t) . \mathrm{e}^{-it} \mathrm{d}t - B(d) - G_0$$
$$+ (G_T - D_T) \mathrm{e}^{-iT}. \quad (7.1)$$

The first-order conditions for a maximum present value obtained by differentiating (7.1) with respect to T and d, and simplifying are:

$$d . p(t) - f(d) . m(t) = i(G_T - D_T) \quad (7.2)$$

$$B'(d) = \int_0^T p(t) . \mathrm{e}^{-it} \mathrm{d}t - f'(d) \int_0^T m(t) . \mathrm{e}^{-it} \mathrm{d}t. \quad (7.3)$$

The first of these conditions (7.2) states that, for an optimal life, the building should be demolished when the net annual income from the property is just equal to the income which could be obtained from the investment of the net proceeds of the sale of the land after deducting the costs of demolition. The second condition (7.3) states that, at the optimal density, the present value of the net income to the developer of an additional space unit is just equal to the cost of constructing it.

The effects of the leasehold system

The analysis above can be used to investigate theoretically the possible effects of the leasehold system on the development.

> Building leases appear in various forms but their basis is that the landlord lets a site to a tenant for a long period, usually 99 years, at a ground-rent, and the tenant is obliged to erect a building which then reverts to the landlord at the end of the lease. The theory is that the building has then reached the end of its economic life and that it will be demolished and replaced by the same tenant or his successor-in-title (McDonald, 1969, p. 179).

Obviously, if the length of the lease and the optimal life of the building really are the same when a site is developed at its optimal density, then the site is developed in exactly the same way as it would if the freehold were owned by the developer. In this case alone, the system of land tenure makes no difference; but if the length of the lease differs from the optimal life of the building, then the length of the lease may be taken into account by the developer as a constraint affecting both the density of development and the life of the building to be erected. One would intuitively expect that he would attempt to make the expected economic life of the building as close as possible to the life of the lease.

On the other hand, at the beginning of the lease, when it has 99 years to run, the length of the lease is not going to be one of the developer's major problems. Firstly, the returns expected 99 years in the future will be very uncertain; hence, the optimal life of the building will itself be uncertain. Secondly, the present value of the net income from the project in 99 years' time will be very small indeed in relation to the present value of the income from the project in the first few years of its life. Decisions about the optimal density will be much more important than decisions about the optimal expected life. For example, if a development yields a net income of £1,000 throughout its 99-year life, a 1 per cent increase in density will increase the present value of the development by 1 per cent, while a 1 per cent increase in the life increases the present value of the property by about 0·04 per cent. Thirdly, the developer won't be alive when the lease expires. If the development is carried out by a company, the manager won't be alive either. Errors in judgement will therefore affect neither the developer's wealth nor the manager's salary.† For these reasons we would expect that the length of the lease has little effect on the expected life and density of the development carried out.

The term of the lease may become a problem only towards the end of its life. The building on the site may have an optimal life which is either more or less than the term of the lease. If the optimal life of the building under normal conditions is longer than the term of the lease the developer can attempt to maximise the return which he obtains in the unexpired term by reducing the value of the property which reverts to the ground landlord. He can do this by spending on repairs and maintenance only what is necessary to prolong its life up to the end of the lease. If he doesn't have to reduce rent charges by as much as he reduces expenditure on maintenance, his net income from the property will be increased at the expense of the ground landlord.

If the optimal life of the building is less than the term of the lease the developer has three alternatives. Firstly, he can attempt either to negotiate a new lease or to sell (or give back) the unexpired term of the lease to the ground landlord. In most circumstances it would be to the ground landlord's advantage to agree either to purchase or negotiate since, in either case, he sells a new lease for a further term. However, it may not always be to his advantage. In particular, the ground landlord may own the freehold of several adjacent properties on which the leases are all due to expire on the same date. If the ground landlord intends that the area should be redeveloped as a whole, then he will prefer to wait rather than accept the unexpired term.

In this event, the developer can attempt to run the property for as long as the net income after paying maintenance make some contribution towards the payment of ground-rent, thus prolonging the life of the building beyond that originally

---

† Weingartner (1969) suggests that one reason for the popularity of 'the payback period' as an investment criterion is that, unlike the net present value method used here, it allows managerial judgements to be checked after the project has been running only a short time.

expected. Once the expenditure on maintenance threatens to exceed the income from rents the developer can vacate the building for the duration of the lease.†

The developer's third alternative, if the unexpired term of the lease is long enough, is to demolish the existing building and erect on the site a temporary building with an expected economic life approximately equal to the unexpired term of the lease.

It can be seen that the effects of the leasehold system on patterns of location are unpredictable. We shall therefore ignore it in the remainder of the book.

CAPITAL VALUES – 2

The mathematical model used in the preceding section was adequate for the formulation and analysis of the investment decision which must be made by the developer at the time of the construction of the building. At this time some approximate estimate of the life of the building must be made and the optimal density must be decided. In the analysis of an investment project it is plausible to assume that the income and expenditure of the entrepreneur over the life of the project to be, can at least be estimated. In the analysis of the effect of a leasehold system, however, the model was being pushed further than was really correct, given its assumptions. At one point in the discussion we stated that the developer could reduce his maintenance expenditure, and implied that this might cause a reduction in his rent income. This statement, though plausible, is not implicit in the mathematical model either as an assumption or a conclusion. It is stated there that rents are a function of

† Property taxes have been ignored. Under the British rating system, rates are generally levied only if the building is occupied. If rates are taken into account the life of the building would be prolonged only until the costs of repairs and maintenance, plus the expenditure on rates, exceeds the rental income. Under a site-value rating system, tax would be payable whether or not the building were occupied. The building would therefore be let for as long as the net income from rents, less the cost of repairs and maintenance, made some contribution to the ground-rent and property tax. The building would therefore be let for a shorter period under a system where rates are not levied on unoccupied property than under a system of site-value taxation, or no taxation at all.

time, and the cost of repairs and maintenance is a function of time, but it is not assumed that rents are a function of maintenance.

Nevertheless, it is obviously true that rents do depend on maintenance expenditure, so that for the analysis of the effect on an existing property of changing circumstances we shall construct a dynamic model incorporating this assumption as an alternative to the earlier purely static analysis of the investment decision.

We assume that the developer wishes to maximise his profits over the life of the building, T. At any time, t, the developer will own a certain stock of capital, $y(t)$, represented by the property. With this capital, y, and at that particular time, t, the developer can take certain decisions as to repairs, maintenance, renovation, improvements, subdivision, etc. of the property. We denote these decisions by $x(t)$. The property at any time can be let at a rent, r, which is a function of the stock of capital, $y(t)$, i.e. $r = r(y(t))$.†‡ What we wish to do is to determine the characteristics of the time path of the decision variable, x, which leads to the maximisation of the developer's profits, i.e we wish to determine the developer's optimal maintenance policy. As set out above, the problem is quite clearly a problem in optimal-control theory (see Dorfman (1969), for example).

We assume that the developer wishes to maximise the present value, π, of his profits over the life of the building, and:

$$\pi = \int_0^T (r(y(t)) - x(t))e^{-it}dt \qquad (7.4)$$

where the rate of interest is $100i$ per cent. The rate of change of the capital invested in the building is equal to the amount spent on maintenance per unit time less the depreciation of the capital. Thus,

$$y'(t) = x(t) - \delta.y(t) \qquad (7.5)$$

† Note that we are discussing r, the rent of the whole property, not p, the rent per space unit.

‡ I am indebted to R. M. Solow for pointing out that, if r were a function of t as well as of y, the conclusions from the analysis would be incorrect and the optimal capital investment would vary over time. In my view, this would be true if there is capital obsolescence which – unlike physical deterioration – cannot be countered by 'maintenance' or expenditure on modernisation. If it can, then the analysis is still correct.

where δ is the rate per unit of time at which a unit of physical capital deteriorates.

Let $\psi(t)$ be an auxiliary variable associated with the marginal value of a unit of capital, and $\psi'(t)$ be the rate of change over time of the marginal value of a unit of capital. The first-order conditions for a maximum† are (7.5), together with:

$$e^{-it} = \psi(t) \qquad (7.6)$$

$$r_y . e^{-it} - \psi(t) . \delta = -\psi'(t). \qquad (7.7)$$

These three conditions can be interpreted thus: (7.5) states how capital grows at any instant as a result of the decisions made with respect to improvement and maintenance and of the total capital at the time; (7.6) states that the marginal value of a unit of capital at time t is the present value of that unit of capital; (7.7) states that the rate of depreciation of the marginal value of a unit of capital at time t is equal to the present value of the marginal increase in output at time t less the value of the physical depreciation of the unit of capital.

Differentiating (7.6) with respect to t, we obtain:

$$-i . e^{-it} = \psi'(t). \qquad (7.8)$$

By substituting for $\psi(t)$ and $\psi'(t)$ in (7.7), using (7.6) and (7.8), and simplifying, we obtain:

$$r_y = i + \delta. \qquad (7.9)$$

This is the equation defining the optimal path for profit maximisation. It states that, along such a path, the rate of maintenance and improvement at each time r must be chosen so that the marginal increase in rental income, during any short interval of time, resulting from an additional unit of capital must be just sufficient to cover the cost of borrowing the unit of capital and the rate of physical deterioration of capital.

Equations (7.9) and (7.5) define both the optimal rate of maintenance and improvement and the optinal amount of capital at any point in time. From (7.9) we know that capital must be invested in the property up to the point at which $r_y = i + \delta$. This defines a unique amount of capital. But if the amount of capital is fixed it follows that $y'(t) = 0$, and hence,

† Apart from the transversality conditions.

from (7.5), that the optimal rate of maintenance is that which just compensates for the physical deterioration of capital.

The situation can be illustrated graphically. In Fig. 7.6 the capital represented by the property is shown on the vertical axis, and the rate of expenditure on improvement and maintenance is shown on the horizontal axis. The optimal capital is indicated in the diagram by the horizontal straight line AB. At a capital of OA, equation (7.9) holds. Line OC is the graph of the equation $x - \delta.y = 0$. Any point on this line represents a combination of maintenance and capital such that the expenditure on maintenance compensates for the rate of physical deterioration of capital. Any point not on this line denotes a combination of maintenance and capital such that capital is either being added to, because improvement and maintenance more than compensates for physical deterioration, or being reduced, because maintenance fails to compensate for deterioration. Thus, at point D, capital is decreasing, as indicated by the direction of the arrow, while at point E, capital is increasing. The only stable point, under the given conditions, is therefore point F, at which capital is optimal and remains optimal.

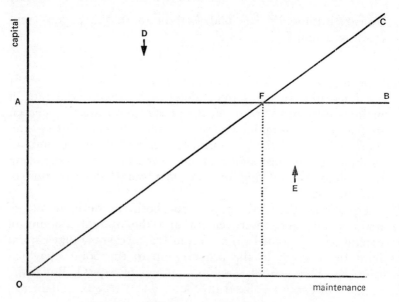

Fig. 7.6 The optimal rate of maintenance.

It can be seen that provided neither the demand conditions, (affecting r_y), nor the rate of physical deterioration, δ, nor the rate of interest, i, change over time the condition of the property will remain the same. But what happens if any of these parameters change. Changes in the rate of interest are not of particular interest, since we cannot assume that changes will systematically affect the housing market, but changes in demand conditions may depend on location, and the rate of physical deterioration may vary with the age of the property. These two parameters will therefore be discussed in turn.

Suppose that δ does increase over time. It would follow from (7.9) that the optimal amount of capital is less, since it can be assumed that a lower capital will yield a higher marginal-rent product. In Fig. 7.7 this is shown by the new horizontal line $A'B'$. The slope of the diagonal line OC will also change, since at each level of capital a higher rate of expenditure on maintenance would be required to make up for the faster rate of deterioration. The intersection of the lines OC' and $A'B'$

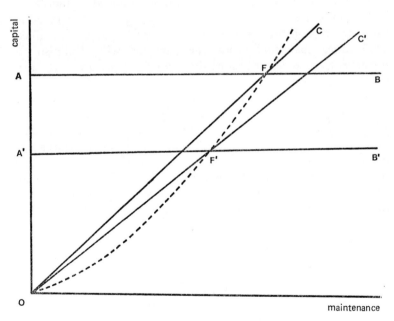

Fig. 7.7   The effect of changes in the rate of deterioration.

at F' indicates the new optimal rate of expenditure on maintenance.

The locus of the point showing the optimal capital and rate of expenditure on maintenance as δ changes, is shown by the dotted line $OF'F$. From (7.9) we know that $\delta = r_y - i$ at the optimum. Substituting in (7.5), we obtain:

$$y'(t) = x(t) - (r_y - i)y(t). \qquad (7.10)$$

At the equilibrium point, at which $y'(t) = 0$, it follows that (7.10) describes the equation

$$x = (r_y - i)y$$

or

$$y = \frac{1}{r_y - i} \cdot x.$$

This is the equation of curve $OF'F$ in Fig. 7.7. $OF'F$ slopes upward at an increasing rate, since $r_y > i$ as $\delta > 0$, and it is plausible to assume that r_y diminishes as capital increases. From the slope of the curve it follows that as δ increases, so not only the optimal capital decreases but the optimal rate of expenditure on maintenance decreases. In effect, over time, the quality of the housing will diminish as it deteriorates physically and the amount spent on its upkeep diminishes.

Suppose now that δ remains constant but the demand conditions for housing change at the location of the property. We will assume that, whereas the location was formerly one where people with low incomes lived, it now changes status and becomes an optimal location for people with high incomes. It is to be expected that the increase in rents obtained from an additional unit of capital, at any level of capital, will be higher. That is to say, the new higher-income residents will be willing to pay more for improvements than the old residents were. Hence, at any y, r_y is now greater, and it follows that the optimal level of capital investment defined by equation (7.9) is increased. Hence, with the shift in the status of the area as an optimal location, the houses will be improved and the rate of expenditure on maintenance increased.

It can also be seen that the reverse follows. If an area

becomes an optimal location for low-income families, after previously being an optimal location for high-income families, then r_y will be lower at all levels of y. The optimal policy for the landlord is therefore to disinvest in the property by reducing maintenance expenditure and thus allowing the property to deteriorate.

The filtering process

In the literature of land economics it is frequently stated that one of the features of the market for housing is the 'filtering process'. Admittedly, as Lowry (1960) and Grigsby (1963) have shown, there has been some disagreement about the precise definition of filtering, but the received theory would appear to be that when new houses are built for the highest-income families, the houses which they vacate would become available to the families in the next to highest-income group. And so on, with houses filtering down the income scale, and families filtering up the housing scale. Implicit in the theory is the supposition that the quality of housing is correlated with its age, so that the highest-income groups occupy the newest, highest-quality housing and the lowest-income groups occupy the oldest, lowest-quality housing.

This filtering process is also the basis of the theories of residential location suggested by Burgess and Hoyt and discussed in Chapter 1. The distribution of income groups within the city is supposed to be determined by the growth of the city, which results in the newest houses being at the edge and the oldest being near the centre.

The theoretical analysis above suggests that, even though the quality of housing may to some extent be determined by its age, another, possibly more important, determinant of housing quality will be the income of the households occupying it. Thus, although high-quality housing is associated with high-income families, and low-quality housing with low-income families, the theoretical analysis suggests that the quality of the housing is determined by the incomes of the residents rather than the residents being determined by the quality of the housing. This becomes relevant in Chapter 8 when we discuss the pattern of location of different income groups in the inner city.

Depreciation in the value of housing

In the theoretical analysis of the effect of variations in the rate of physical deterioration, δ, over time, we left open the question as to whether, in fact, it does vary. It is certainly true that empirical studies show that the value of a given property tends to decline over time. For example, in Chapter 5 it was shown that the market value of flats on long leases appeared to decline by about 4 per cent per annum. This agreed with the results quoted by Grebler, Blank, and Winnick (1956) which showed that the value of a property appeared to decline by about 4 per cent per annum in the first few years of its life. On the other hand, their study also shows that the rate of depreciation falls off rapidly after that, becoming much less in the later years of the life.

Of course, depreciation in value may occur because of changes in occupancy. As Burgess observed, the poorer families tend to be located in the inner areas of cities in the older housing. But, as shown above, we would expect that housing occupied by poor families would be badly maintained, and if it had formerly been occupied by higher-income families we would expect its value to have fallen. Hence, the observed depreciation in the value of housing may be due, at least in part, to factors associated with locational change instead of an increasing rate of physical deterioration.

A second problem occurs because of the way in which expenditure on maintenance may fall due. In the specification of our model we assumed that the rate of physical deterioration may be a constant rate per annum. This may well be true but it may not be possible to spend money on maintenance at a constant rate, because of indivisibilities. Thus, to cite an obvious example, the house may need repainting only every three or four years though the paintwork deteriorates at a constant rate. This will probably be true of most items of expenditure on maintenance; they will only fall due every few years.

For the first few years of a building's life none of these items of maintenance expenditure will be payable, even though the building fabric may be deteriorating. The buyer of a new house should have to spend little on maintenance for some time. The buyer of an old house may have to spend a great deal on maintenance. It follows that a new house will sell for a higher price than an old house, even though they may be otherwise identical.

Suppose, therefore, that a house is maintained as far as possible in the same condition over its life, and can thus be let at a constant rent. The gross rental income over time is shown by the horizontal line *IR* in Fig. 7.8. Suppose the expenditure on maintenance increases fairly sharply after the first few years and then levels off. The net income from the property over time would then be that indicated by the line *II*.

We might expect that a graph of the value of the property as a function of time would have the same shape as *II*, but this can be shown to be incorrect. The value of the property at any time will be equal to the present value of the net income from the property over the rest of its expected life. This value will depend on the rate of interest which is expected to prevail. Obviously, this rate of interest must lie between zero and infinity. Fig. 7.9 shows the shapes of the graphs of present value as a function of time, in the case of each of the extreme assumptions. For the sake of comparison the values are stated in terms of an index where the initial value of the property (i.e. its value at the date of construction) is equal to 100.

When the rate of interest is infinitely high the present value

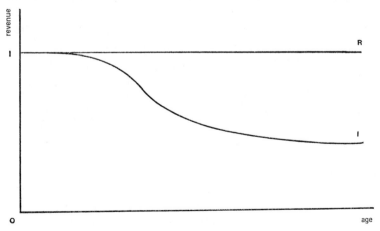

Fig. 7.8   Constant gross rent but declining net income.

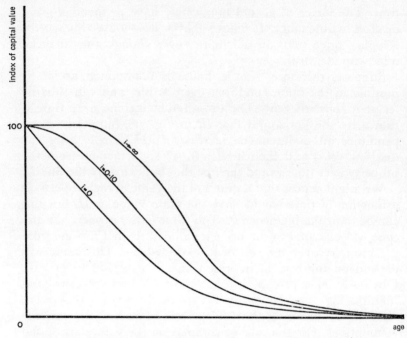

Fig. 7.9   Capital value as a function of time and the rate of interest.

of the property in any year will be equal to the net income in that year. The present value of the rent in all future years will be equal to zero. In this case, and in this case alone, the shape of the graph of rent as a function of age will be exactly the same as the shape of the graph of value as a function of age, and it is shown as the upper curve in Fig. 7.9.

When the rate of interest is zero the present value of the property will be equal to the total net income from the property up to the end of its life. The net income in each year will be valued equally. It follows that the present value of the net income from the property at the end of the first year of its life will be less than its value at the beginning. The value at the beginning of the year will include the value of the net income in that year; the value at the end of the year will exclude it. The value at the end of the first year will therefore be less than the value at the beginning by an amount equal to the net rent in that year. Similarly, the present value at the

end of each year will be less than the value at the beginning
of the year by an amount equal to the net income in that year.
It follows that, if net income in each consecutive year is
constant, the present value of the property will fall at a constant
rate per annum equal to the net income. Therefore, at the beginning of the property's life, when expenditure on maintenance
is negligible, the value of the property will fall, as shown by
the lower curve in Fig. 7.9. If the rent from the property
declines over time, the rate at which the value of the property
falls will also fall. Overall, one would expect that the graph
of present value would have approximately the shape of the
lower curve in Fig. 7.9, provided that the graph of net income
had the shape shown in Fig. 7.8. The value of the property
will decline at a faster rate in the early years, but at a slower
rate thereafter.

Obviously, the rate of interest in reality will lie somewhere
between the extreme values of zero and infinity, but will
obviously be closer to zero than infinity. We would therefore
expect that the graph of the value of a property as a function
of time will have a shape closer to that of the lower curve than
the upper curve in Fig. 7.9. The empirical evidence is in
agreement with this result.

Two conclusions follow from this analysis. Firstly, the rate
of physical deterioration may be constant, but the rate of
capital depreciation may be higher in the early years of the
life of a building. Secondly, it is perfectly possible that a property
can be rented for a constant amount over its life while its
capital value depreciates. The first of these conclusions shows
that the empirical evidence of capital depreciation is perfectly
compatible with a theoretical analysis which assumes a constant
rate of physical deterioration. The second shows that empirical
evidence of capital depreciation is perfectly compatible with
the assumption that the market rent of a property remains
constant.

Thus, the empirical evidence of capital depreciation does
not necessarily support the view that the rate of physical
deterioration of capital increases over time. Indeed, the fact
that the rate of capital depreciation decreases over time tends
to confirm the hypothesis that the rate of physical deterioration
is constant. Furthermore, if the rate of physical deterioration

increased over time, it would follow that buildings would have a definite life. But as Grebler, Blank, and Winnick (1956) state:

> experience [does not] indicate that a specified finite physical life can be assigned to residential structures. . . . Rather, the life of a structure is indefinitely long and is terminated much more frequently by casualty or site obsolescence than by physical deterioration or structure obsolescence (p. 377f).

Hence, we shall assume that the rate of physical deterioration is constant and that variations in housing quality can be assumed to be caused by variations in the demand conditions.

CONCLUSIONS

Of the two topics discussed in this chapter, only the second may be supposed to be relevant to the pattern of residential location in the city. The variations in density discussed in the first section of the chapter, though an important topic, cannot be expected to make a systematic difference to the pattern of residential location. In the chapters which follow, we use the theory that has been developed to predict the patterns of residential location and we present evidence confirming these predictions.

The present chapter has, therefore, been something of a digression from the main argument of this book, but was necessary to make a start on tying up various loose ends in the theory of the supply of space. Much further empirical and theoretical work needs to be done in this field but this is not the place for it. In Chapter 8 we return to the theory of residential location and attempt to analyse the patterns of location of different income groups.

8 Patterns of Location – I: Economic Status

In this chapter we use the theory of location developed in Chapter 3, to explain the way in which the distance from the centre at which a household finds its optimal location varies with its income, and hence, the existence of concentric zones differentiated by household income. Then, by modifying the theory to take account of the fact that 'individuals *prefer* to interact with others who are socially similar to themselves' (Anderson, 1962), we explain the existence of sectors differentiated by household income. We also show that the sectoral pattern is likely to exist in smaller cities while a pattern of concentric zones is more likely to occur in a large city. Finally, we explain the pattern of location of high-income households in the central city and the way in which this pattern may change over time.

Throughout the chapter we assume that differences in income between households are uncorrelated with other differences which might alter the optimal locations of households, so that in analysing the pattern resulting from differences in incomes we can ignore any other differences between households. In the next chapter we explain the way in which the household's optimal location may vary with its stage in the family cycle. Empirical support for the assumption that income and stage in the family cycle are uncorrelated is provided by the numerous factor analyses of the 'human ecology' of Western cities which almost invariably define two (uncorrelated) factors which are identified with 'economic status' and 'family status'. As Abu-Lughod (1969) has pointed out, however, the two may not be uncorrelated in non-Western cities, and the explanation presented here might therefore have to be modified to be applicable to non-Western cities.

INCOME AND DISTANCE FROM THE CITY CENTRE

In Chapter 3 we found that the necessary condition for the household's location to be optimal is that the variation in total travel costs caused by a small move away from that location either towards or away from the city centre will be just equal to the variation in rent. Formally:

$$q.p'(t) + c'(t) + r.v'(t) = 0 \qquad (8.1)$$

where q denotes the number of space units occupied by the household, $p'(t)$ denotes the rate at which the rent per space unit declines with time spent travelling from the city centre, $c'(t)$ denotes the rate at which the direct (financial) cost of travel increases with time spent travelling, r denotes the rate of pay of the head of the household (who is assumed to be the only member of the household who is gainfully employed), and $v'(t)$ is a positive fraction such that $r.v'(t)$ indicates the household's marginal valuation of his travel time (per hour) as a fraction of his rate of pay.

Each of the functions $p(t)$, $c(t)$ and $v(t)$ can be decomposed to read $p(k(t))$, $c(k(t))$, and $v(k(t))$, thus stating p, c, and v as functions of distance where distance is a function of time. Then (8.1) can be written in the form:

$$q.p_k.k_t + c_k.k_t + r.v_k.k_t = 0$$

where, for example, $p_k.k_t = p'(k(t)).k'(t)$. We can then cancel k_t throughout to obtain:

$$q.p_k + c_k + r.v_k = 0 \qquad (8.2)$$

where the necessary condition for an optimal location is expressed in terms of distance instead of time spent in travelling.

Equation (8.1) can also be written in the form:

$$p'(t) = -\frac{c'(t) + r.v'(t)}{q} \qquad (8.3)$$

and (8.2) can be written in the form:

$$p_k = -\frac{c_k + r.v_k}{q}. \qquad (8.4)$$

In this form, (8.4) gives the slope of the rent gradient at the optimal location and (8.3) gives the rate of change of the rent per space unit as a function of time spent in travelling. Obviously, (8.4) could be obtained from (8.3) by dividing through by k_t, the speed of travel.

We know that (8.4) also gives the slope of the lowest attainable bid-price curve, p_k^*, for

$$p_k^* = - \frac{c_k + r.v_k}{q} \qquad (8.5)$$

since, at the optimal location, the slope of the lowest attainable bid-price curve is equal to the slope of the rent gradient. We could also assume the existence of a set of bid-price curves which would describe the rent per space unit as a function of time rather than distance, and the slope of the lowest attainable bid-price curve of this type, p_t^*, would be given by (8.3), i.e.

$$p_t^* = p_k^*.k_t = - \frac{c'(t) + r.v'(t)}{q}. \qquad (8.6)$$

Again (8.5) could be obtained from (8.6) by dividing through by k_t, as (8.6) could be obtained from (8.5) by multiplying through by k_t.

Since the slope of one type of bid-price curve is a multiple of the slope of the other type, we can use either type of bid-price curve in any analysis. The characteristics of one type of bid-price curve (rent as a function of distance) were described in Chapter 6, and this description applies *pari passu* to the other type (rent as a function of time).

Suppose that the above equations describe the conditions at the optimal location of a particular household k^* miles and t^* minutes from the city centre paying a rent per space unit of p^*. How does the household's optimal location change if, for some reason, its demand for space increases? This increase in its demand for space results in the formation of a new set of bid-price curves, and these bid-price curves will be less steep at all points. Thus, the slope of the new bid-price curve passing through the point (k^*, p^*), or (t^*, p^*), will be less steep than the slope of the old curve, since the denominator q of the fractions indicating the slope in (8.5) or (8.6) will have increased, but the numerator will remain the same since the cost of travel remains the same. The household will therefore have to move further out from the centre to find its optimal location. Less technically, we can say that the increase in the space required by the household means that the total rent savings which would result from a move away from the centre become greater, while the travel costs remain the same, so that the household gains from a move away from the centre.

On the other hand, how does the household's optimal location change if, for some reason, the rate of pay of the head of the household increases but its demand for space remains the same? The slope of the new bid-price curve passing through the point (k^*, p^*), or (t^*, p^*), will be steeper than the slope of the old curve, since the numerator of the fractions indicating the slope in (8.5) or (8.6) will have increased because of the increase in r but the denominator q is assumed to remain the same. The household will therefore move nearer to the centre to find its optimal location. Less technically, the increase in the householder's rate of pay, and the resultant increase in his valuation of his travel time, means that the total saving in travel costs which would result from a move closer to the centre are increased, while the increase in total rents remains the same, so that the household gains from a move towards the centre.

It is, however, implausible to assume that, if the householder's rate of pay were to increase, his demand for space would remain the same. It is to be expected that the quantity of space demanded by the household would also increase. Therefore, when the householder's rate of pay is increased, two opposing forces work to change his location. The first, the increase in the rate of pay, pulls the household's location towards the centre because of the increase in the value of the household's travel time. The second, the increase in the household's demand for space, pushes the household away from the centre.

The direction of the change in the household's location, if any, will depend on the relative strengths of these two opposing forces. If the household's demand for space increases very little when the rate of pay is increased, the household will move towards the centre; if the demand for space increases very greatly when the rate of pay is increased, the household will move out from the centre. The direction of the move can be said to depend on the household's elasticity of demand for space with respect to income. Now it is obvious that there is some elasticity at which the two opposing forces are exactly equal, so that any increase in the householder's rate of pay would not result in a move either away from or towards the centre (though he would either enlarge his house or move to a larger house the same distance from the centre). But the householder's optimal location will only remain the same if the new bid-price curve at the point (k^*, p^*), or (t^*, p^*), after the increase in the rate of pay has exactly the same slope as the old bid-price curve at that point. Then both the old and the new curves will be tangental to the rent gradient at that point, and the household's optimal location will be exactly k^* miles and t^* minutes from the centre, both before and after the pay increase. Suppose that the rate of pay of the head of household increases by $\triangle r$ to $r + \triangle r$, and the quantity of space which he would occupy k^* miles (t^* minutes) from the centre at a price p^* increases in consequence by $\triangle q$ to $q + \triangle q$. Then the household's optimal location will remain the same if the slope of the two lowest attainable bid-price curves are the same, or:

$$-\frac{c'(t) + (r + \triangle r)v'(t)}{q + \triangle q} = -\frac{c'(t) + r.v'(t)}{q} = p_k^* k_t. \qquad (8.7)$$

Multiplying out and simplifying, reduces this to a condition that, as $\triangle q$ and $\triangle r$ tend to zero

$$\frac{\partial q}{\partial r} = \frac{q.v'(t)}{c'(t) + r.v'(t)}.$$

If the household's elasticity of demand with respect to the rate of pay (or income elasticity) is denoted by ϵ, the household's optimal location will remain the same if

$$\epsilon = \frac{r}{q} \cdot \frac{\partial q}{\partial r} = \frac{r}{q} \cdot \frac{q.v'(t)}{c'(t) + r.v'(t)}$$

$$= \frac{r.v'(t)}{c'(t) + r.v'(t)}. \qquad (8.8)$$

Alternatively, since $v'(t) = v_k.k_t$ and $c'(t) = c_k.k_t$, we can divide through by k_t, and state that the household's optimal location will remain the same if

$$\epsilon = \frac{r.v_k}{c_k + r.v_k}.$$

If the elasticity at which the household's optimal location remains the same is denoted by E, then, if the household's income elasticity of demand for space is greater than E, the household will move outward since the outward push generated by the increase in the demand for space will outweigh the inward pull generated by the increase in the value of travelling time. On the other hand, if the household's income elasticity is less that E, the household will move inward since the inward pull generated by the increased travel costs will outweigh the outward push of the increased demand for space.†

Thus we cannot predict the pattern of location of households with different incomes in relation both to each other and to the city centre without some knowledge of the income elasticity of demand for space of the households. The patterns which result when elasticities are high differ from those which result when elasticities are low, and both patterns from those which result when elasticities are at intermediate levels.

It can be seen from (8.8) or (8.9) that the income elasticity, E, at which the optimal location remains the same when the householder's rate of pay changes is a function of the rate of pay, i.e.

† The effects of changes in the price of space due to changes in location can be ignored: either (1) the household's optimal location remains the same, so that the price of space also remains the same and the quantity of space demanded is not affected by variation in price; or (2) the household moves outward, the price of space falls, and the household's demand for space increases, thus reinforcing the outward push of reduced rents (the quantity of space demanded must increase with the change in price since the substitution effect is necessarily positive, and the income effect is also known to be positive because the household's income elasticity of demand for space must be positive for it to be moving outward); or (3) the household moves inward and the increase in price leads to a reduction in the quantity of space demanded, thus reinforcing the strength of the inward pull of reduced journey time (provided it can be assumed that the household's income elasticity is greater than 0, so that both the substitution effect and the income effect are positive).

$$E(r) = \frac{r \cdot v'(t)}{c'(t) + r \cdot v'(t)} = \frac{r \cdot v_k}{c_k + r \cdot v_k}.$$

The graph of this function E is drawn in Fig. 8.1. It is assumed that the values of v_k, k_t, and c_k are known parameters. The graph of E shows that $E(r)$ increases at a diminishing rate as r increases, tending asymptotically towards one from below as r tends to infinity.

The simplest case occurs when all households have income elasticities greater than one, since the income elasticities are then necessarily greater that $E(r)$. For every household an increase in the rate of pay must cause a move to a new optimal location further from the centre of the city. The population will therefore distribute itself so that the poorest households with the smallest demands for space are located closest to the centre of the city and the richest households with the largest space demands are located at the periphery. This pattern of location will occur in any city in which all the households have income elasticities greater than $E(r)$, whether or not the elasticities are also greater than one.

At the other extreme, if the income elasticities of all the households are very low, so that over the whole range of rates of pay and the heads of household the income elasticity is less than $E(r)$, then, for every household, an increase in the

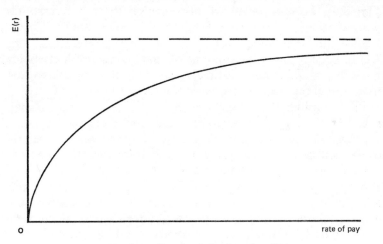

Fig. 8.1   Graph of the function E.

householder's rate of pay must result in a move to a new optimal location closer to the centre of the city. The population will therefore distribute itself within the city so that wealthy households are located adjacent to the city centre while poor households are located at the periphery.

Note that the residential patterns in the two extreme cases described above correspond with the accepted images of the residential patterns in the North American city and the Latin American city. Of course, these are images of simplified patterns. In Latin America, in the last half century, there has been a tendency for 'the earlier relationship between social status and residential location with reference to the centre . . . to be *reversed* in the course of time and to assume the "North American pattern" ' (Schnore, 1965, p. 358). A possible explanation of this change will be given in Chapter 10. Again, the pattern in the North American city is more complex than the accepted image, with many of the richest households located at the centre. This is just the sort of pattern which might occur in the case when some households have income elasticities which are less than $E(r)$ while some have elasticities which are greater than $E(r)$, as we shall now show.

Suppose that some households have the same income elasticity ϵ'. The rates of pay of the heads of these households range between r_1 and r_3, and $\epsilon' = E(r_2)$, where $r_1 < r_2 < r_3$ (see Fig. 8.2). For simplicity we also assume that all households with a rate of pay of r_1 would occupy the same amount of space (hence all households with a rate of pay r, where $r_1 \leq r \leq r_3$, would occupy the same amount of space, since all households have the same income elasticity) and that all households value their travelling time at the same fraction of the rate of pay.

First consider those householders for whom the rate of pay r is less than r_2. For each of these households the income elasticity ϵ' is greater than $E(r)$ so that any increase in the rate of pay will lead to a move to an optimal location further from the city centre. These households will distribute themselves so that those paid at the rate r_1 will locate nearest the centre, and those paid almost r_2 will locate furthest from the centre. Now consider those whose rate of pay is greater than r_2. Since these households have an income elasticity ϵ' which is always less than $E(r)$, they will distribute themselves so that those house-

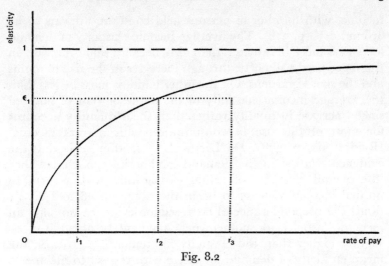

Fig. 8.2

holds paid at the rate r_3 will locate nearest the centre, and those paid just over r_2 will locate furthest from the centre. Those paid at a rate r_2 will not, of course, change locations on a change in pay, and it can be seen that they will locate furthest from the centre of all the households with an income elasticity ϵ'. These results suggest a mixed pattern with the intermediate-income group paid at r_2 at the edge of the city, and both the very rich and the very poor located towards the centre of the city. The pattern of location is as shown in Fig. 8.3.

It can be assumed that in the modern British or American city the income elasticities of the households will be dispersed over a wide range about the mean, though even the mean income elasticity is difficult to assess. Margaret Reid (1962), in her exhaustive study of *Housing and Income* in the U.S.A., found that 'the income elasticity of rooms with respect to normal

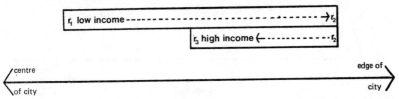

Fig. 8.3   Location of income groups relative to each other and to the city centre (income elasticity of demand for space equal to ϵ').

income, with number of persons held constant, appears to be around 0·5' (p. 378). The average income elasticity of demand for space must be greater than this because some increase in the space used will occur through increases in the size of rooms and the size of gardens. On the other hand, it must be less than the average income elasticity of the value or rent of housing since 'increase in quality rather than sheer quantity accounts for most of the rise in consumption with normal income' (Reid, 1962, p. 378). De Leeuw (1971), after reviewing the evidence relating to the demand for housing, concluded that 'the overall elasticity of rental expenditure with respect to normal income appears to be in the range of 0·8 to 1·0 . . . [and] the preponderance of cross-section evidence supports an income elasticity for homeowners moderately above 1·0, or slightly higher than the elasticity for renters' (p. 10). So the mean elasticity of demand for space with respect to income in the U.S.A. (and probably also in Britain) must lie between 0·5 and 1·0 and the income elasticities of households will be dispersed about this average value.†

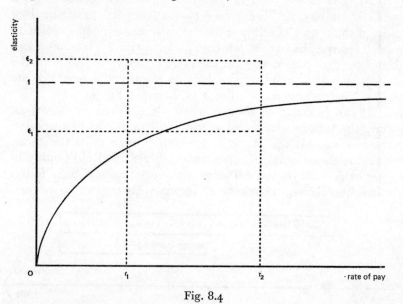

Fig. 8.4

† Byatt, Holmans, and Laidler (1972) estimate the income elasticity of the (value of) housing to lie between 0·7 and 1·0 for Britain.

Suppose that the income elasticities of the households in a city range between ϵ_1 and ϵ_2, that the rates of pay range between r_1 and r_2, and that $\epsilon_1 < E(r_2)$ and $\epsilon_2 > E(r_1)$. This is illustrated in Fig. 8.4, from which it can be seen that the income elasticities of those households with the highest elasticities are greater than $E(r)$ over the whole range of incomes. As shown above, they will locate so that the poorest are nearest the centre, and the richest nearest the edge. On the other hand, the income elasticities of the other households will not always be greater than $E(r)$. The income elasticities of the wealthier households will be less than $E(r)$. The pattern of location of the households with an income elasticity less than $E(r)$ at high-income levels will be as described above, with the richest and the poorest locating towards the centre, and a middle-income group nearest the periphery (see Fig. 8.3).

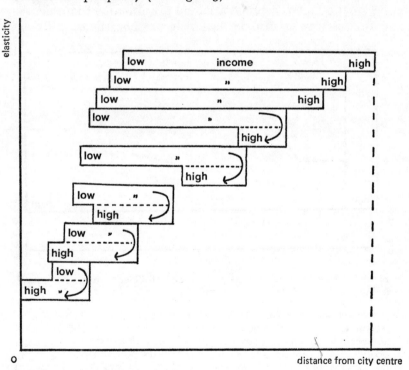

Fig. 8.5 The distribution of income groups in the city relative to each other and to the city centre.

What will be the pattern of location of *all* households of the city, allowing for both variations in income and in income elasticity? A possible distribution of the population of households relative to each other and to the city centre is shown in Fig. 8.5, and in a different way, in Fig. 8.6. Figure 8.5 shows the distribution of the population at each of a number of income elasticities. Figure 8.6 shows the proportion of the population in each income group as the distance from the centre increases. Note that all that is illustrated is a possible ordering of the income groups, relative to the centre and to each other, which is based on an assumption about the dispersion of income elasticities about the mean. If we wanted to know the proportion of the total area occupied by each income group we should have to know something about the distribution of the population over the range of incomes.† For example, Beckmann (1969) assumes a Pareto distribution of income.

In this case it can be seen that the households with the

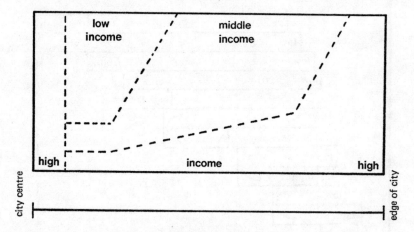

Fig. 8.6   The proportion of the population in different income groups and its variation with distance from the centre.

† We should also have to know something about the distribution of income from property. The receipt of property income would increase a household's demand for space, but the valuation of travel time would not necessarily increase by as much as it would if the income were received in the form of wages. It follows that households receiving property income would locate further from the centre than they would if all their income was in the form of wages.

highest rates of pay are the exclusive occupants of the residential
area adjacent to the city centre, just as they are the exclusive
occupants of the peripheral areas of the city. The households
with the lowest rates of pay locate near the centre in the ring

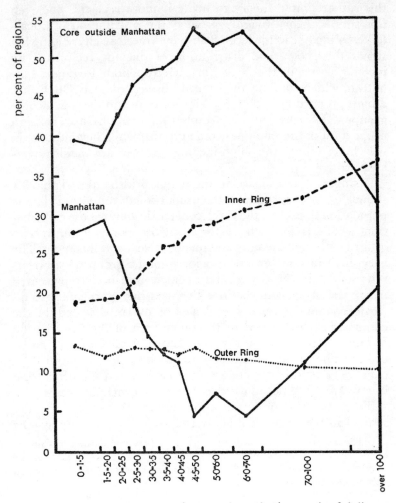

Fig. 8.7   Geographic distribution of income recipients by income classes,
New York metropolitan region, 1949.
'Income recipients' means families and unrelated individuals reporting
income.) *Source:* Hoover and Vernon (1959).

adjacent to the high-income households. The average rate of pay is lowest in this ring, and it increases from this low point as distance from the centre increases.

In the next section of the chapter we will attempt to modify this rather simple theory to make it more realistic. But even this simple theory roughly predicts the distribution of households by income in the large British or American city. Figure 8.7 shows the geographic distribution of 'income recipients' by income classes in the New York Metropolitan Region. Manhattan, which contains the central business district (CBD), has a high proportion of both the high-income and the low-income groups. The core outside Manhattan contains a very high proportion of the middle-income groups, while beyond this, in the inner ring, the high-income groups are again over-represented.

A similar pattern exists in the major British cities. Table 8.1 shows, for six conurbations, the proportion of the occupied male population in each of the three concentric zones of the conurbation which is in each of three socioeconomic groups: two groups with high incomes and one with very low incomes.† The very high-income groups (socioeconomic group 1) is over-represented in the conurbation centre, and underrepresented in the rest of the conurbation. The next highest income groups (socioeconomic groups 1 and 4) are overrepresented in the conurbation centre and in the outer area of the conurbation but underrepresented in the central city outside the conurbation centre. The lowest income group (socioeconomic group 8) is overrepresented in the conurbation centre, and the proportion in any area declines as distance from the centre increases.

These patterns of location are similar to the patterns described in the theoretical analysis above. Since we can assume

† The data on incomes is obtained from the National Readership Survey of the Institute of Practitioners in Advertising – from a probability sample of adults aged 16 or over in Great Britain interviewed during the period July to December 1960:

	Mean gross income of the head of household
Socioeconomic group	
3 Professional Self-Employed	£1,730
4 Professional Employees	£1,130
1 Employers and Managers – large organisations	£1,110
11 Unskilled manual	£ 420

TABLE 8.1

Geographical Distribution of Socioeconomic Groups in the British Conurbations, 1961.

	Group 3			Groups 1 and 4			Group 11		
	Conurbation centre (%)	Rest of central city (%)	Rest of conurbation (%)	Conurbation centre (%)	Rest of central city (%)	Rest of conurbation (%)	Conurbation centre (%)	Rest of central city (%)	Rest of conurbation (%)
Greater London	2·1	0·7	0·9	9·4	5·7	10·2	11·1	11·2	6·4
West Midlands	1·9	0·5	0·5	7·9	5·0	7·2	13·1	8·1	7·2
South-east Lancashire	3·3	0·5	0·7	5·9	4·5	7·2	23·3	10·9	9·3
Merseyside	2·9	0·6	0·6	7·4	4·9	7·8	22·4	15·1	12·1
Tyneside	3·8	0·7	0·5	5·3	6·2	6·2	18·0	11·8	11·3
Clydeside	0·7	0·5	0·9	4·3	3·8	5·9	20·7	12·8	11·1

Sources: Census 1961, England and Wales: Socioeconomic Group Tables, Table 1 (General Register Office, London, 1966). Census 1961, Scotland: Occupation, Industry, and Workplace Tables, Part I; Occupation Tables, Table 2 (General Register Office, Edinburgh, 1966). Census 1961, Scotland: Occupation and Industry Tables, County Tables, Glasgow and Lanark, Table 5 (General Register Office, Edinburgh, 1966).

that the (unknown) mean and dispersion of the household income elasticities are approximately the same in all six cities, the similarities in the distribution of the population in the cities, to some extent, confirm the theoretical analysis. Variations in the geographical distribution seem to result from variations in the income distribution between the conurbations, with the population of London having a higher average income than that of the other conurbations.† Note, however, that, as first stated in Chapter 2, because of systematic variations in residential density with distance from the centre, the patterns of location found in these cities, though predicted by the theory, may not be the best attainable for society as a whole. The theory is positive and not normative.

SOCIAL AGGLOMERATION

The above theoretical analysis explains the location of the different income groups in concentric zones about the city centre. But Hoyt's studies of the location of high-income neighbourhoods in American cities have shown that the geographical distribution of income groups may be sectoral. In this section of the chapter we shall modify the simple theory presented above in order to explain this sectoral pattern. The modified theory shows that a mainly sectoral pattern is to be expected in the small town and that, the larger the city, the more likely it is that the pattern will be one of concentric zones.

The basic assumption which we make is that people will generally try to live in the same neighbourhood as others in the same socioeconomic groups. We will use the term social agglomeration to describe this grouping together so that we can draw attention to the similarity to industrial agglomeration in Weberian location theory.‡

There are several reasons for this grouping together, which we could call the economies of social agglomeration. Firstly, people in different income groups spend their money in different ways. The shops selling to people with high incomes carry a

† A possible explanation of this difference in income distributions is given by Evans (1972b).

‡ The terminology is due to Marian Bowley who also drew my attention to the parallel.

stock of goods different to that carried by shops selling to people with low incomes. Households will therefore be able to reduce the cost of their shopping trips by living in the same neighbourhood as others in the same income group, since the shops in the neighbourhood will carry the sort of goods they wish to buy. Furthermore, specialist shops and consumer services may find it profitable to open in neighbourhoods where all the residents are possible customers, but would not exist if their customers were widely scattered since their sales would be reduced and they would not be profitable. These specialist shops and services will make an area even more attractive to those in the right income group. Also the public services e.g. school quality, desired by people in different income groups will vary. If people locate in the same political unit as others in the same income group they can get the services they wish.

Secondly, as Anderson (1962) has noted, 'many researches have shown that interaction rates are higher among individuals within a subgroup than among individuals between subgroups. On the basis of this evidence it is reasonable to assume that individuals *prefer* to interact with others who are socially similar to themselves'. If people in a given income group prefer to interact with others in the same group it is plausible to assume that they will prefer to live in the same area as the others and that, *ceteris paribus*, the greater the number in the same group in that area, the greater the advantage of living in that area will be.

The assumption that social agglomeration will occur can also be justified using the concept of social space. The household's 'neighbourhood space' is determined by the network of relationships 'that encompasses daily and local movement' (Buttimer, 1969, p. 420). If the household prefers to interact with others in the same income group its neighbourhood space will be smallest in area if it locates in the same neighbourhood as others in the same income group. Buttimer notes that 'Chombart de Lauwe has calculated thresholds in space beyond which certain groups cannot travel without experiencing frustrations, tensions, and feelings of anomie' (p. 421).†

† But it should be noted that de Lauwe's researches related only to the Parisian working class, and his results may not be generally applicable.

Now it can be assumed that each household would wish to ensure that, within its neighbourhood space (i.e. in its daily or local movement), it did not have to cross this threshold. It can only do this by locating in the same area as others in the same social group and income group) where shopping and contact with friends (i.e. daily or local travel) does not require it to make lengthy journeys.

The result of social agglomeration will be a sectoral pattern of location of different income groups, and this pattern will vary with the size of the city. In a small town, instead of distributing themselves in a thin ring around the perimeter of the city, the high-income group will collect together in a neighbourhood in one sector towards the edge of the city. Moreover, if the number of households in the high-income group who might otherwise wish to locate at the centre or in intermediate areas is small, these households may find that together they cannot create a viable neighbourhood. If the radius of the city is small, they may choose to locate in the high-income neighbourhood at the periphery, giving up the advantages of proximity to the centre to gain a location in a desirable neighbourhood.

As the size of the city increases so the number of people in the high-income group who wish to live near the centre increases, and a viable neighbourhood becomes feasible. Moreover, as the size of the city increases, so the distance from the centre to the edge of the city increases, and so the cost of giving up proximity to the centre increases relative to the benefit of living in the high-income neighbourhood at the periphery. For both these reasons, the large city is likely to have a high-income neighbourhood at, or near, the centre, while in a small town all members of the high-income group will locate in the outer part of a sector of the city.

We will now show that the shape of the high-income neighbourhood at the edge of the city is likely to vary with the size of the city, with the sector tending to become a ring as the city increases in size. An informal argument can be stated briefly. As the city increases in size the number of people in the high-income group also increases, and so the high-income neighbourhood increases in area. If the neighbourhood retained a circular shape as it expanded, those living on the outer edge of the neighbourhood and the city would find themselves incurring higher and higher travel costs. They could reduce these costs by moving away from the outer edge of the neighbourhood to its side, where they would still be on the edge of the neighbourhood. They will therefore do this, and the neighbourhood will therefore expand circumferentially rather than radially, tending towards a ring as the city increases in size.

The more formal and rigorous argument is lengthier and more abstract. We assume that members of the high income group are willing to pay a higher rent for a location amongst other members of the group.† We could also assume that members of other income groups would be willing to pay higher rents for location among others in the same income group, but members of the high-income group will probably pay the highest premiums so that it seems plausible to concentrate on the analysis of the behaviour of this group. There is some empirical evidence that the value of land is higher in high-income areas. Stone (1964) notes that the value of land in London falls regularly with distance from the centre, but that 'the steadiness of the fall is disrupted by the distribution of areas with special attractions as residential locations'. As an example of such a location, he cites Hampstead in the high-income sector of London. Muth (1969) from a study of housing in six U.S. cities, found that, 'if anything, the intensity of housing output per unit of land is greater in higher-income neighborhoods . . . if the effects of differences in location are removed' (p. 202), which would imply that the value of land is also higher.

We can assume that the 'neighbourhood premium' which a high-income household would be willing to pay to locate in a

---

† It would be possible to derive the same conclusions by assuming that households are only willing to pay a negligible premium to locate in the same neighbourhood as others in the same income group and that they do not notice small 'errors in location'. From this, it would follow that the ncrease in location costs which would result from location up to a certain distance from the optimal location relative to the city centre would be so small that it could be ignored. Thus, only a slight preference for location with others in the same group would still result in a sectoral pattern.

An assumption similar to this is made by Devetoglou (1965). The reason for not using this assumption is that, as noted in the text, there is empirical evidence to support the other assumption.

high-income neighbourhood will vary with the location within the neighbourhood, being highest at its centre and diminishing to zero at the edge. The variation in the premium is illustrated in Fig. 8.8 which shows a cross-section of the rent gradient along the radius of the city passing through the centre of the high-income sector. For simplicity, we assume that all the members of the high-income group have similar tastes and wish to locate at the edge of the city. The bid-price curve which would determine the location of the high-income group in the absence of any neighbourhood effects is indicated by the dotted line *QP*. The members of the group bid up the price of space at some site on the periphery in their desire to locate together. This site may have topographical advantages, as Hoyt suggests, in that it may be on high ground or on a river or lake front. Otherwise, it may be a matter of chance that a site becomes the centre of a high-income neighbourhood. Thus, Hoyt mentions that some high-income neighbourhoods appear to have grown up about the site of the early home of an

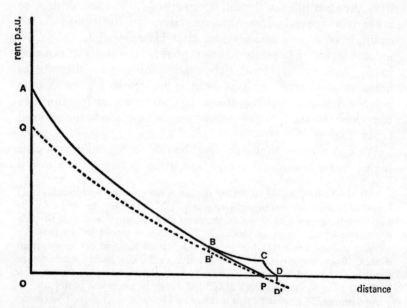

Fig. 8.8 A cross-section through the rent gradient and the high-income neighbourhood.

influential citizen. The high-income sector may also grow up on a fast radial transport route, as we shall show in Chapter 10, but here we still assume for simplicity that radial transport routes are uniform in speed and cost.

In Fig. 8.8 the centre of the neighbourhood is indicated by point C on the rent gradient where the neighbourhood premium is £PC per space unit. At the inner and outer edges of the neighbourhood (points B and D on the rent gradient) the premium is smaller and is indicated by the vertical distance between the bid-price curve and the rent gradient (BB' and DD'). The neighbourhood is shown in plan form by the area $BEDF$ in Fig. 8.9. It can be shown that, given the very simple assumptions made here, the neighbourhood must have this elongated shape. The neighbourhood premium paid by those

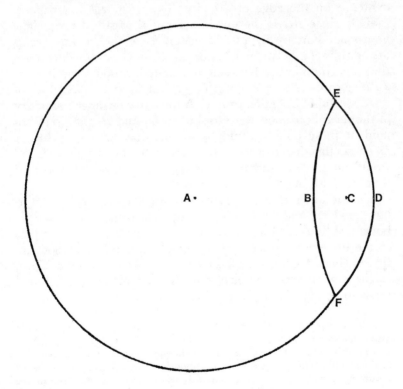

Fig. 8.9   A plan of the city and the high-income neighbourhood.

located at E and F is clearly zero. It follows that if members of the income group are to be indifferent between location and B and D and E and F, then E and F must be further away from the centre of the neighbourhood than B or D is, since a premium is necessarily paid for location at B or D, as can be seen from Fig. 8.8.†

Suppose that £PC is the maximum premium that anyone will pay. They are willing to pay this when a neighbourhood has a population and area large enough for the neighbourhood space of a person located at C to be just enclosed by the neighbourhood. We assume that $BEDF$ is exactly this minimum optimum size. It can be shown that, if the population of the city increases and the same income distribution is maintained, the neighbourhood will either expand circumferentially (i.e. along the axis EF) or new neighbourhoods will come into existence on the edge of the city. The area will not expand radially since BD is the maximum radial width of the neighbourhood. For, by hypothesis, members of the income group are indifferent between location at B or C or D. Therefore, they are indifferent between a neighbourhood premium of £BB' payable BC miles from C, £PC at C, and a premium of £DD' payable CD miles from C. If the outer or inner boundary of the neighbourhood were to shift to D_1 and B_1 the premium paid at these points would be greater than that payable at D or B, the distance from the centre of the neighbourhood would be greater, and people living there would be worse off than at C, all of which contradicts the hypothesis.

The radial width is therefore fixed at BD miles. As the neighbourhood expands circumferentially, the centre of the neighbourhood becomes the arc of a circle, as do the middle sections of the inner and outer boundaries of the neighbourhood, all three being concentric circles centred on the city centre. Thus, as the city increases in size, the high-income neighbourhood tends to change from a sector into a circle.

---

† Some members of the high-income group would locate in the inner areas and not on the edge of the city if there were no neighbourhood effect. If a neighbourhood existed they would therefore locate on the inner edge of the neighbourhood since their 'neighbourhood premium' would then be reduced. It can easily be seen that this would result in the area of the high-income neighbourhood becoming more 'sectoral' in shape.

Obviously, the theoretical analysis set out above is based on a very simplified model of a city. Social factors which might affect the geographical distribution of the high-income group are not considered. In particular, we shall show in Chapter 14 that a sectoral patteern of location by income groups may arise because the different income groups are employed at different, spatially separated workplaces in the city, instead of all being employed at a central workplace.

Despite its limitations however, the analysis does explain the empirical results obtained in recent statistical analyses of the spatial variation of economic status in several U.S. cities.

> If we compare the between-sector and between-zone variances in a number of different sized cities for which we have analyses of variance, some interesting though very tentative findings emerge . . . There appears to be a regular ordering, by size of city, of the relative contributions of [concentric] zones and sectors to the variation in socio-economic status . . . For the larger the city the greater the importance of zonal variation of socio-economic status as compared with sectoral variation, although sectoral variation remains the more important in all the cities (Rees, 1970b, p. 373).†

The tendency for the importance of zonal variation to increase relative to sectoral variation as the city increases in size is even noticeable in one of the maps published in Hoyt's original study. Figure 8.10, reproduced from Hoyt (1939, p. 115), shows the 'shifts in location of fashionable residential areas in six American cities' between 1900 and 1936. In each case the sectoral pattern has changed to become more zonal.

It is to be expected that sectoral variation will always remain important, no matter how large the city. As the earlier theoretical analysis showed in Fig. 8.6, in addition to the high-income households who locate on the periphery or adjacent to the city centre, there will be many who find their optimal location in the intermediate areas of the city. These households will group together in neighbourhoods at various distances from the centre. They will attain maximum benefit from this

† The ordering was first noted by Anderson and Egeland (1961).

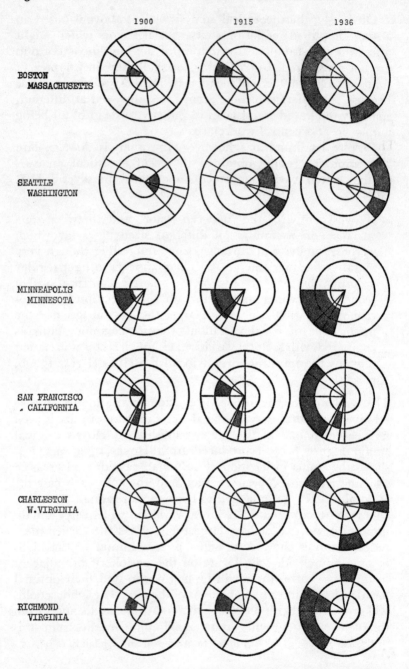

Fig. 8.10 Shifts in location of fashionable residential areas in six American cities, 1900–1936. Fashionable residential areas are indicated by shading. *Source:* Hoyt (1939).

grouping if these neighbourhoods lie in a single sector of the city so that they coalesce. This will lead to the creation of high-income sectors even when the city is very large; examples are the North Shore in Chicago and the Hampstead/Highgate sector in London.

THE CENTRAL CITY

The high-income households who locate, at, or near, the city centre provide one of the most interesting problems in the analysis of patterns of residential location. For one thing, many American authors seem to imply that no such households exist or that, if they do, they are at a suboptimal location. Muth (1969) appears to argue that, since the *average* income elasticity of demand for space is high, *all* households will respond to an increase in income by moving further out. But even though the average is high, *some* households may have low income elasticities and will therefore respond to an increase in income by moving further in. In analysing the geographical distribution of households by income, we are interested in the dispersion of elasticities, as well as the average. If the dispersion is wide, many households will have low elasticities, even when the average is high, and so some households with high incomes and low elasticities will wish to locate at the centre.

When the city is large enough, the number of high-income households wishing to locate at the centre will be enough to make a high-income neighbourhood viable at the centre. These households will not ring the CBD but will form the neighbourhood on one side of the CBD, so that the households will have the advantage of proximity to each other. Thus, in large cities we would expect to find – and do find – high-income areas bounded on one side by the CBD, and on the other by very low-income areas.

Given that these high-income areas exist in the central cities, what happens when, for some reason, the size of the high-income group wishing to locate at the centre increases? The neighbourhood can expand either into the area of the CBD or, radially and circumferentially, into the low-income areas. The first alternative is likely to be impossible if the city is growing, since the area of the CBD will then also be increasing and may even be encroaching on the area of the high-income neighbour-hood. Thus, the only practicable alternative is for the high-income households to move radially and laterally into space in the low-income areas. Until the neighbourhood reaches its optimal width, both radial and lateral movement is likely. When it reaches its minimum optimal width the movement will be lateral, turning the neighbourhood from a sector to a ring.

We demonstrated in Chapter 7, firstly, that housing in low-income areas is likely to be in a run-down condition, since the amount spent on maintenance will be low; and secondly, that high-income households will wish to live in well-maintained housing. Therefore, if a high-income household moves into housing in a low-income area it will wish to renovate the property. Now it has been argued by Davis and Whinston (1961) that a badly maintained property is unlikely to be renovated when surrounded by property which is similarly maintained. They argue that the main influence on the value of a property is not the condition of the property itself but the condition of the property in the surrounding area. If a house is renovated its value will increase only slightly because the adjacent properties will still be run down. It will not therefore be worth while for property to be renovated in these circumstances since the owner will not fully recover any capital he invests in the house.

While the plausibility of this argument must be admitted, it leaves unexplained the facts that property in low-income areas does get renovated, and high-income households do move into low-income areas. Admittedly, this process is not well documented, but it takes place. Gillian Tindall (1971), in *New Society*, portrays a stage in the transition of a street in London from low-income to high-income occupation, with half the houses being owned by high-income households. Michael Frayn (1967), in his novel *Towards the End of the Morning*, treats the process as being well known. It has now even acquired a name – 'gentrification'.

The most plausible reconciliation of the Davis–Whinston argument with the evidence appears to be this. The high-income households who are willing to move into low-income areas and renovate property are not headed by executives to whom the prestige of an address is important, but by architects, journalists, TV producers, etc., for whom the house alone is important. By renovating a house in a run-down area they will be able to obtain proximity to the centre at a reasonable cost. Furthermore, because the prestige of the address is unimportant to them, they will not regard the value of the house as significantly diminished by its surroundings. They may even feel that they will gain status with their friends by living in such an area, so demonstrating their lack of snobbishness.† The businessman, on the other hand, may feel that if he lived in a low-income area, even in renovated housing, it would demonstrate his lack of financial success and cause him to lose status with his friends.

As more and more houses in the area change hands and are renovated, the status of the area rises and it becomes permissible for business executives to move in. The value of those houses which have already been renovated therefore increases. Moreover, as the rise in the status of the area is anticipated, the unrenovated property increases in value, and the low-income households – the original inhabitants – are forced out by increasing rents. In due course the area is only occupied by high-income households and becomes recognised as a high status area. Moreover, with the increase in the value of space in the area it becomes worth while to redevelop the original properties to a high density.

Thus, the Davis–Whinston argument is only partly correct. The quality of the surroundings will affect the value of a house but they will not necessarily prevent its renovation. Of course, it may be that several factors peculiar to Britain make renovation much more likely in British cities. The government

† Frayn, in his satirical novel, describing the motives of Dyson and his wife writes: 'They decided to find a cheap Georgian or Regency house in some down-at-heel district near the centre. However depressed the district, if it was Georgian or Regency, and reasonably central, it would soon be colonised by the middle classes. In this way they would secure an attractive and potentially fashionable house in the heart of London, at a price they could afford; be given credit by their friends for going to live among the working classes; acquire very shortly congenial middle-class neighbours of a similarly adventurous and intellectual outlook to themselves; and see their investment undergo a satisfactory and reassuring rise in value in the process'.

grants paid to the owners of houses which are improved by the installation of certain basic amenities almost certainly encourage high-income households to buy and renovate old housing. The fact that any profit made from the sale of one's home is exempt from capital-gains tax gives an extra incentive to owner occupiers to seek out houses which will appreciate in value.† It is possible that the green belts, around London in particular, encourage high-income households to locate in the inner area of the cities by preventing the continuous expansion of the cities and disrupting the markets for high-income housing on the periphery of the built-up areas. It has also been suggested that the intrinsic beauty of the Georgian houses which exist in the low-income areas of inner London, in Islington for example, encourage high-income households to move into these areas, but Fulham, which has mainly Victorian housing and into which high-income households are now moving, shows that Georgian architecture is certainly not necessary for an area to rise in status.

Even in the absence of these special factors, renovation of old housing does take place in the United States, though the only hard evidence which I have been able to find relating to

TABLE 8.2

Survey of Central City Residents, New York, Chicago, and Philadelphia, 1957. The Distribution of Primary Wage Earners by Occupation

	Percentage in each occupational class		
Occupational class	White sample total	Apartment house	Rehabilitated house
I Top executive	13	17	7
II High professional	14	15	12
III Secondary managerial	26	29	23
IV Creative arts	17	10	26
V Low professional	8	7	9
VI Clerical, white-collar	7	7	8
VII Blue-collar, service	2	–	4
VIII Out of labour force	14	15	12

Source: Abu-Lughod (1960), Table A.4.

† I am indebted to D. Higham for this point.

the above hypotheses appears in a sample survey of upper-middle and upper-class residents of the central cities of New York, Chicago, and Philadelphia carried out by Janet Abu-Lughod (1960) in June 1957. (Her Table A.4 is reproduced here as Table 8.2.) It shows 'the percentage of primary wage earners (almost always head of household) in various occupations both in the total sample and in the apartment and rehabilitation samples separately' (p. 402). The titles of the occupational classes are self-explanatory, with the exception of IV, Creative arts. This class includes 'creative and performing artists, writers, artists, musicians, etc., whether working independently, for mass media, or for commercial firms' (p. 402). Abu-Lughod notes that

> the most dramatic difference [in occupational composition of the two sub-samples], and one which is highly significant statistically, is the difference found in category IV, the creative and performing arts. A much higher percentage of persons in this occupational group live in rehabilitated dwellings than in the new apartment houses. From this one may conclude that the 'arty' or quasi-bohemian element of well-to-do center-city residents is attracted almost exclusively to rehabilitated town houses, whereas the new luxury apartment houses attract such occupational groups as executives and high professionals (p. 403).

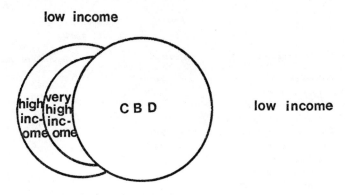

Fig. 8.11 The geographical distribution of income groups in the inner area of the city.

The association between income and occupation should also be taken into account in explaining the movement of households into renovated or rehabilitated housing in low-income areas. If a neighbourhood forms, the very high-income groups would constitute the core of the neighbourhood while the next-highest income groups would occupy its outer periphery. An outline plan of the geographic distribution of income groups in this inner area is shown in Fig. 8.11. Any increase in the population and area of this high-income neighbourhood will push those on its periphery out to the low-income areas. Therefore, those households renovating housing in the low-income areas will have lower incomes than those in the core who live in flats. If those in the 'creative arts' have lower incomes than business executives the difference in the occupations of those living in the two types of housing could arise for this reason alone.†

Partial support for this argument is provided by the data in Table 8.3 (reproduced from Table A.5 in Abu-Lughod (1960)). Within occupation groups those with lower incomes tend to be in rehabilitated houses. Between occupation groups there is

TABLE 8.3

Survey of Central City Residents. New York, Chicago, and Philadelphia, 1957. Households with Incomes of $15,000 and more by Occupation of Primary Wage Earner

	Percentage of households having $15,000 or more income	
	Apartment house	Rehabilitated house
Occupational class		
All households	49	33
I Top executive	90	93
II High professional	58	50
III Secondary managerial	42	40
IV Creative arts	48	30
V Low professional	32	26
VI Clerical, white-collar	0	0
VII Blue collar, service	0	0
VIII Out of labour force	27	5

Source: Abu-Lughod (1960), Table A.5.

† I am indebted to J. T. Eve for this point.

some tendency for those in the lower-income occupations to live in rehabilitated houses. But it can be clearly seen that variations in income do not wholly explain differences in the type of house occupied. These are explained by both income and occupation together.

SUMMARY

In this chapter we have used the theory of household location set out in Chapter 3 to explain the geographical distribution of households with differing incomes. By making the additional assumption that households wish to locate in the same area as households with similar incomes, the theory explained the patterns of location observed in empirical studies. Using the elementary theory of the dynamics of the housing market, which was developed in Chapter 7, the theory also explained the pattern of location of high-income groups in the central city.

In the next chapter we explain the location of the household at different stages in the family cycle.

9 Patterns of Location – II: Family Characteristics

In this chapter the theory of household location is used to predict and explain the patterns of location of households with differing family characteristics, i.e. differences in the number of persons in the household and in the number working. In the first section of the chapter we use the theory to predict the changes in the household's optimal location within the city caused by changes in the household's family characteristics. In the second section these predictions are compared with the results of some empirical studies of intra-urban migration. The third section is mainly concerned with the comparative length of the journey to work of males and females in the British conurbations. The results of several empirical studies of the patterns of location of households with different family characteristics are discussed in the fourth section. In a final section we show that household size varies with distance from the centre.

PATTERNS OF LOCATION

In Chapter 3 we showed that the first-order condition for a household's location to be optimal is that, at that location, the increase or decrease in travel costs (both direct and indirect) caused by a small move towards, or away from, the city centre must be just equal to the decrease or increase in the total rent. Formally:

$$q \cdot p_k + c_k + r \cdot v_k = 0 \qquad (9.1)$$

where q denotes the number of space units occupied by the household, p_k the rate at which the rent per space unit declines with distance from the city centre (i.e. the slope of the rent gradient), c_k the rate at which the direct financial cost of travel increases with distance, r is the rate of pay of the householder, and v_k is a fraction such that $r.v_k$ denotes the householder's valuation of his travel time (per mile) as a fraction of his rate of pay. Equation (9.1) can be written in the form

$$p_k = -\frac{c_k + r.v_k}{q} \tag{9.2}$$

which gives the slope of the rent gradient at the optimal location. In this form the equation also gives the slope of the lowest bid-price curve attainable at the optimal location by the household, i.e.

$$p_k^* = -\frac{c_k + r.v_k}{q} \tag{9.3}$$

Suppose now that we assume that (9.2) and (9.3) describe the first-order condition for the optimal location of a household consisting of one unmarried man, this location being k_1 miles from the city centre with a rent per space unit of p_1. Where relative to this location, would he locate on marriage? Suppose that his wife does not go out to work: if the new household remained at a location k_1, miles from the centre, the direct cost of travel per mile would remain the same, and there is no reason to expect the indirect (time) cost of travel to alter. On the other hand, the new two-person household will require a larger quantity of space than a one-person household. The slope of the bid-price curve of the two-person household through the point (k_1, p_1) will be less steep than the slope of the bid-price curve of the single-person household through (k_1, p_1), since the numerators of the fractions indicating the slope will be the same but the denominator will be larger. The argument shows that, at any location and at any price per space unit, the slope of the bid-price curve of the two-person household (one person working) will be less steep than the slope of the bid-price curve of the one-person household. In Chapter 6 we showed that of any two households the one with the steeper bid-price curves will locate nearer the city centre. Hence, a man will locate his home closer to the city centre if he is single than he would if he were married and his wife didn't go out to work.

This argument also shows that every increase in the number of children in the household moves the optimal location of the household further from the centre. For with every increase in the number of children the quantity of space demanded by the household increases but the cost of travel (both direct and indirect) remains the same. Hence, the slope of the household's bid-price curve decreases and its optimal location moves further from the centre.

Suppose now that we consider the location of the married couple when both husband and wife go out to work. Its optimal location will almost certainly be closer to the centre than the optimal locations of the man and woman if they lived in separate single-person households. The total amount of space required by two persons living together will almost certainly be less than the total amount of space they would require living separately. Living together, they can share facilities such as a kitchen or bathroom which would otherwise be necessary for each. Suppose that the direct and indirect cost of travel is the same for each partner and that their demands for space are the same. Before marriage, each of the partners will locate at the same distance from the centre k_1, paying a rent per space unit p_1 for a quantity of space q_1. Suppose then that we treat the two persons living separately as constituting a single 'household'; the slope of the lowest attainable bid-price curve of this household will be:

$$p_k^* = -\frac{2c_k + 2r.v_k}{2q_1}.$$

The slope of the new household's bid-price curve through the point (k_1, p_1) will be steeper than the slope of the bid-price curve(s) of the old 'household' (the two living separately), since the numerator of the fraction indicating the slope will be the same (for the cost of travel does not alter) but the denominator will be less than $2q_1$. The same argument can be used to show that, at any location and at any rent per space unit, the slope of the bid-price curve of the household formed by the two persons living together will be steeper than the slope of the bid-price curves of each person living separately if their tastes are the same. Hence, the home of the couple living together will be closer to the city centre than were their homes when they lived separately.

Similar arguments can be used to show that a group of three

persons, all having similar tastes and working for similar rates of pay, would live closer to the centre of the city if they shared living accommodation than if they each lived separately, or even if two shared and one lived alone. Every increase in the number of workers sharing living accommodation moves their optimal location closer to the centre. When a very large number of workers share accommodation in a hostel, its optimal location is very close to the centre indeed.

Obviously, the conclusions of the above argument depend to some extent on the assumption that all those sharing accommodation have similar tastes and are paid at similar rates, but even if the optimal locations of the partners (if they lived separately) were at different distances from the centre, it can still be seen that the location of the shared accommodation would be closer to the centre than the previous location of at least one of the partners, and might be closer to the centre than the previous locations of all of them.

In sum, the analysis suggests that the location of households with different family characteristics relative to the centre will be as shown in Fig. 9.1.

Since we have not taken full account of variations in income, Fig. 9.1 really only indicates the pattern of location of households where those at work are paid at a similar rate of pay.

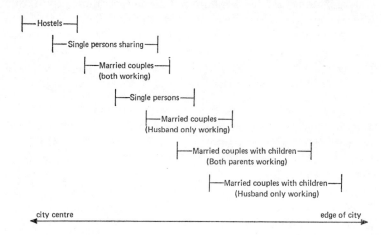

Fig. 9.1 The relative location of households with differing family characteristics.

Yet, since the same pattern of location relative to the city centre can be predicted for the workers and their households for each rate of pay, we can make some general predictions.

1. Within each income group the proportion of the population who are below working age increases with distance from the centre; therefore, for the population of the city as a whole, the proportion who are below working age will increase with distance from the city centre

2. Within each income group the proportion of the adult female population who work outside the home declines with distance from the centre: therefore, for the city as a whole, the proportion of the adult female population who work outside the home will decline with distance from the centre.

Taken together, these two predictions imply that the proportion of the population working will decline with distance from the city centre. For example, Fig. 9.2, drawn using 1961 Census data, clearly shows that the proportion of the population which is economically active declines with distance from the centre of London.

3. Within each income group those females who are at work will tend to be in households where more than one person is working and will therefore be located closer to the city centre than the males who are in work who are more likely to live in households when only one person is working; hence, females working at the centre will have a shorter journey to work than males working at the centre.

With less certainty we can also predict that average household size will increase with distance from the city centre. We have shown that the larger the number of children, the further from the centre is the household's optimal location. But we have also shown that the larger the proportion of workers in a household, the closer to the centre it will locate. If, in general, large households have a much higher proportion of their members going out to work than small families have, then the optimal location of large households would tend to be closer to the city centre

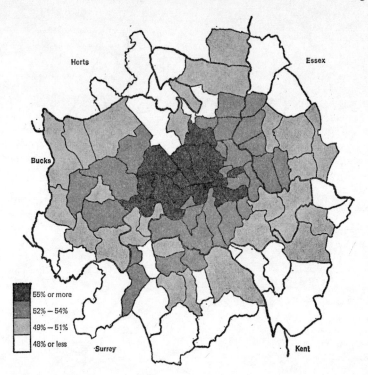

Herts

Essex

Bucks

55% or more

52% – 54%

49% – 51%

48% or less

Surrey

Kent

Fig. 9.2 Proportion of the resident population at work in the Greater London conurbation 1961: proportion of the population at work in local-authority areas. *Source:* 1961 Census of England and Wales.

than would the optimal location of small households, and one might find that the average household size decreased with distance from the centre.

The prediction therefore depends on the rate of decline in the proportion of workers per household with distance from the centre being low relative to the rate of decline in the proportion of population at work. Mathematically, since:

$$\frac{P}{H} = \frac{W}{H} \bigg/ \frac{W}{P}$$

where P is the total population of an area of the city, H is the number of households in that area, W is the number of workers,

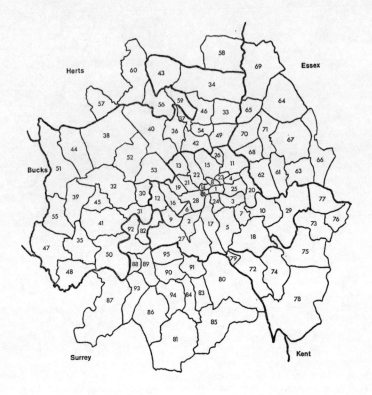

Fig. 9.3   Greater London conurbation, 1961

Key

London
1 City of London
2 Battersea
3 Bermondsey
4 Bethnal Green
5 Camberwell

6 Chelsea
7 Deptford
8 Finsbury
9 Fulham
10 Greenwich
11 Hackney

12 Hammersmith
13 Hampstead
14 Holborn
15 Islington
16 Kensington
17 Lambeth
18 Lewisham
19 Paddington
20 Poplar
21 St. Marylebone
22 St. Pancras
23 Shoreditch
24 Southwark
25 Stepney
26 Stoke Newington
27 Wandsworth
28 Westminster
29 Woolwich

Middlesex
30 Acton
31 Brentford and Chiswick
32 Ealing
33 Edmonton
34 Enfield
35 Feltham
36 Finchley
37 Friern Barnet
38 Harrow
39 Hayes and Harlington
40 Hendon
41 Heston and Isleworth
42 Hornsey
43 Potters Bar
44 Ruislip-Northwood
45 Southall
46 Southgate
47 Staines
48 Sunbury-on-Thames
49 Tottenham
50 Twickenham
51 Uxbridge
52 Wembley
53 Willesden
54 Wood Green
55 Yiewsley and West Drayton

Hertfordshire
56 Barnet
57 Bushey
58 Cheshunt
59 East Barnet
60 Elstree

Essex
61 East Ham
62 West Ham
63 Barking
64 Chigwell
65 Chingford
66 Dagenham
67 Ilford
68 Leyton
69 Waltham Holy Cross
70 Walthamstow
71 Wanstead and Woodford

Kent
72 Beckenham
73 Bexley
74 Bromley
75 Chislehurst and Sidcup
76 Crayford
77 Erith
78 Orpington
79 Penge

Surrey
80 Croydon
81 Banstead
82 Barnes
83 Beddington and Wallington
84 Carshalton
85 Coulsdon and Purley
86 Epsom and Ewell
87 Esher
88 Kingston
89 Malden and Coombe
90 Merton and Morden
91 Mitcham
92 Richmond
93 Surbiton
94 Sutton and Cheam
95 Wimbledon

* Approximate centre of London.

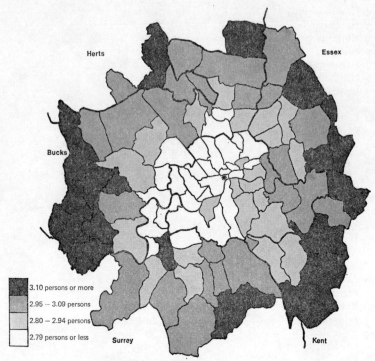

Fig. 9.4   Average household size in the Greater London conurbation, 1961:
average household size in local-authority areas. *Source:* 1961 Census of
England and Wales.

and P/H, W/H and W/P are all functions of distance, it follows
that:

$$\left(\frac{P}{H}\right)_k = \left[\frac{W}{P}\cdot\left(\frac{W}{H}\right)_k - \frac{W}{H}\cdot\left(\frac{W}{P}\right)_k\right]\bigg/\left(\frac{W}{P}\right)^2$$

and since $(W/H)_k < 0$ and $(W/P)_k < 0$, then $(P/H)_k > 0$ if, and
only if,

$$\left[\frac{W}{P}\cdot\left(\frac{W}{H}\right)_k - \frac{W}{H}\cdot\left(\frac{W}{P}\right)_k\right] > 0$$

or

$$\left(\frac{W}{H}\right)_k > \frac{P}{H}\left(\frac{W}{P}\right)_k.$$

In London the rate of decline in the number of workers per
household must be relatively low, and as Fig. 9.2 shows, the

rate of decline in the proportion of the population which is economically active is relatively high. As a result, household size increases with distance from the centre, as shown in Fig. 9.4.† Later in this chapter, and in Chapter 14, we will use statistical techniques to analyse variations in household size in London and in other cities.

INTRA-URBAN MIGRATION

Embedded in the above theoretical analysis is a theory of intra-urban migration over the life cycle. It is implied that the young adult, after leaving the parental home, will live in the inner area of the city, either alone or sharing accommodation with others. On marriage, he and his wife will remain in the inner area, particularly if his wife continues working. With the birth of the first child, and the probability that his wife simultaneously quits the labour force, the forces tying the household to the central area diminish sharply and the household moves out to the inner or outer suburbs, according to its income. There it remains until the children leave home or begin to work and/or the wife starts to work again outside the home. These changes may cause the couple to move back again to the central area until retirement.

What evidence is there to confirm this analysis? There are two questions of interest. First, do households move at the stage of the life cycle indicated by the theory? Second, do they move in the directions predicted? The first question is answered affirmatively by extensive studies of household mobility by geographers and sociologists. In a recent survey article, Simmons (1968), considering the question 'Who moves?', states that U.S. data shows that the major factor causing differential mobility rates is the

life cycle stage [see Fig. 9.5] which obscures the effects of economic level or culture. Annual intra-county mobility rates for children under five are high, about 20 per cent.

† Note that the calculations on which Fig. 9.2 is based include those living in hostels, etc., while those on which Fig. 9.4 is based only include those living in private households. It is not possible to calculate workers per household from Census data.

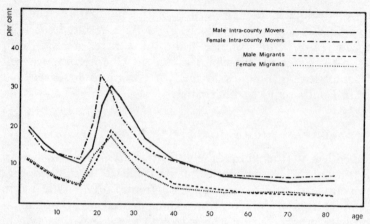

Fig. 9.5 Annual intra-county mobility rate and migration rate of the United States by age and sex, for the population 1 year and over, March 1965. *Source:* Simmons (1968).

They decline for teenagers, maximise at over 30 per cent for people in the young twenties, and decline again to a rate of less than ten per cent for people over 45 years of age.

The British Census data summarised in Table 9.1 show the same sort of variation with age.

In answering the question 'Why do they move?' Simmons states that

all evidence indicates that the most powerful factor in inducing people to change their residence is change in the set of demographic characteristics called the 'urbanization' or 'life cycle' factor.†

Possibly the most intensive study of 'Why families move', using interviews and follow-up surveys, was carried out by Rossi (1955), who concluded that:

The findings of this study indicate the major function of mobility to be the process by which families adjust their housing to the housing needs that are generated by shifts in family composition that accompany life cycle changes (p. 9).

The variations in mobility rates shown in Fig. 9.5 and Table 9.1 confirm Rossi's findings, and we may conclude that, at least

† For British evidence see Wilkinson and Merry (1965).

TABLE 9.1

Mobility Rates, England and Wales, 1961

Age last Birthday	*Moved in previous year* (%)					
	All migrants		*Within same local-authority area*		*Other*	
	Male	*Female*	*Male*	*Female*	*Male*	*Female*
0– 4	14·6	14·6	7·5	7·5	7·1	7·1
5–14	9·4	9·5	4·7	4·8	4·7	4·7
15–24	16·9	21·3	7·6	9·6	9·3	11·7
25–44	15·4	13·0	7·1	5·9	8·3	7·1
45–64	6·5	6·4	3·4	3·3	3·1	3·1
65 and over	6·3	6·8	3·3	3·7	3·0	3·1

Source: Census 1961, England and Wales: Migration Tables, Tables 2 and 12; (General Register Office, London, 1966).

in Britain and the U.S.A., households do move at the stage of the life cycle expected in the theoretical analysis.

Much less research has been put into answering the second question, 'In what direction do households move?' Indeed, Simmons describes the question as 'almost completely neglected'. Here we shall use a method suggested in the *Report on Housing in Greater London* (the Milner–Holland Report) to give an indication of the pattern of intra-urban migration in three British cities: London, Glasgow, and Edinburgh.

> 1951 and 1961 Census data for age groups give some indication of the complex cross movements of population. By looking at 1951 data for certain age groups and comparing the actual number in these groups with the number we should expect to find had there been no migration or deaths since 1951 we can assess the net effects of migration. For example, there were 347,530 young people in the L.C.C. area aged 10–19 in 1951. By 1961 this age group, now 20–29 had increased to 493,440 as a result of net inward migration during the ten year period. (Ministry of Housing and Local Government, 1965, pp. 61, 63).

The percentage difference between the number in the 10–19 age group in 1951 and the 20–29 age group in 1961 is shown in

Fig. 9.6. It shows clearly the inward migration of young adults to form new households in the inner areas of the conurbation, and to that extent confirms the predictions of the theoretical analysis. There is, however, an alternative explanation for this pattern of migration. It could be argued that the only reason for the location of young people near the centre is the attraction of London's entertainments and social life.

To test the predictions of the theory more thoroughly it must be shown that the same pattern of migration occurs in provincial cities with smaller entertainment centres. 1951 and 1961 Census data for age groups by wards is published for the major Scottish cities, but not for English cities. The method of

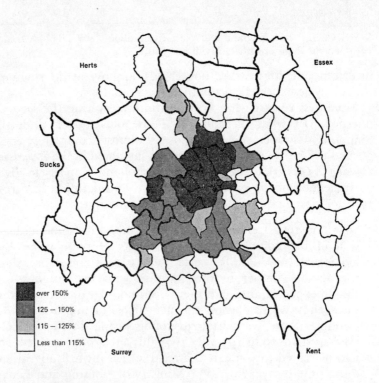

Fig. 9.6 Migration of young people in Greater London, 1951–61. Number in the 20–29 age group in 1961 as a percentage of the number in the 10–19 age group in 1951. *Source:* Ministry of Housing and Local Government (1965).

analysis used in the Milner Holland Report could therefore be applied to Glasgow and Edinburgh. Unfortunately, the geographical redistribution of the populations of the two cities between 1951 and 1961, through slum clearance and the development of new local-authority housing schemes, would swamp the effect of migration of young adults if the same method were used. In the County of London the maximum increase in the population of any borough was 6·2 per cent, and the maximum decrease was 21·1 per cent. In the counties of Glasgow and Edinburgh, on the other hand, the maximum increases in the population of any *ward* were 210·7 per cent and 81·3 per cent, respectively, while the maximum decreases were 37·9 per cent and 30·1 per cent. In these circumstances

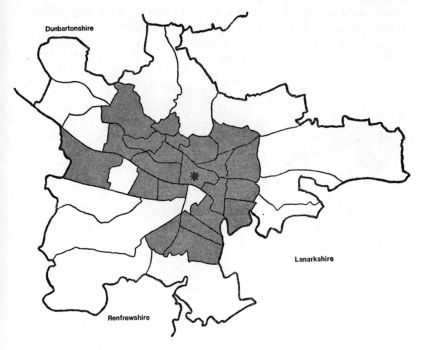

Fig. 9.7   Migration of young people in Glasgow, 1951–61. Wards in which the rate of increase in the number in the group aged 10–19 in 1951 and 20–29 in 1961 was greater than the rate of increase in the population as a whole are shaded. (Asterisk indicates the city centre.) *Source:* Census, 1951, and 1961 County Reports.

the change in the total population of a ward tends to mask the effects of differences in the rates of migration between areas of different age groups. For example, there were 3,263 persons aged 10–19 in the Knightswood ward of Glasgow in 1951. By 1961, the number in this age group, now 20–29, had increased by 1,258 or 38·6 per cent. Yet the total population of the ward had increased by 35,752, or 203·9 per cent between 1951 and 1961. It is obvious that the rate of migration to Knightswood (on the edge of the city) was lower for those aged 10–19 in 1951 than for the other age groups.

Figures 9.7 and 9.8 show hatched those wards of Glasgow and Edinburgh where the rate of increase in the number in the group aged 10–19 in 1951 is greater than the rate of increase in the total population. In this way, allowance is made for changes in the total population of the wards. The maps show those wards where the 20–29 age group were a larger proportion of the total population in 1961 than the 10–19 age group were

Fig. 9.8 Migration of young people in Edinburgh, 1951–61. Wards in which the rate of increase in the number in the group aged 10–19 in 1951 and 20–29 in 1961 was greater than the rate of increase in the population as a whole are shaded. (Asterisk indicates the city centre). *Source:* Census, 1951, and 1961 County Reports.

in 1951. The maps show clearly that in the inner wards of both cities, where the total population usually decreased over the decade, either the rate of decrease in the number of young adults has been less than the rate of decrease of the total population, or the numbers in that age group have increased faster than the rate of increase of the total population. In the outer areas of each city the reverse has generally been the case.

Obviously, it could be argued that the apparent migration patterns in these two cities have resulted from the housing policies of the two corporations, and this is probably part of the truth. But the results for these two cities, taken together with the results for London, certainly do not refute the theoretical analysis. In the next section we shall show that the patterns of location in all major British cities tend to confirm the theory.

LOCATION PATTERNS IN BRITISH CITIES

One of the predictions of the theory is that, since households in which a female is at paid work tend to be households with a high proportion of the members working, these households will locate close to a common work place, and so females will generally have a shorter journey to work. Evidence for the six British conurbations for which a conurbation centre is defined confirms this prediction. Table 9.2 shows that, in each case, the proportion of the females working in the centre and living in either the centre itself or the rest of the central city is higher than the proportion of the males working in the city. Conversely, a higher proportion of males than of females live outside the central city.

The fact that females tend to have a shorter journey to work than males is well known to transport economists, though – according to Clark and Peters (1965) – its discovery dates only from the empirical work of J. F. Kain in the late fifties and early sixties. Kain (1962) showed that, in Detroit, women workers tended to live closer to their work than men. Clark and Peters demonstrate for 17 boroughs in the London region that, in each case, women working in the borough tend to live nearer to the borough than men.

It may be argued that women have a shorter journey to

TABLE 9.2

Proportion of Males and Females employed in the Conurbation Centre and residing in each Subdivision of the Conurbation and outside the Conurbation, Major British Conurbations, 1961

| Central city of conurbation | Six | Subdivision of conurbation | | | | |
		Conurbation centre (%)	Rest of central city (%)	Rest of conurbation (%)	Outside conurbation (%)	Total (%)
London	Male	8·1	34·8	39·1	18·0	100·0
	Female	12·2	44·3	34·3	9·2	100·0
Birmingham	Male	3·0	63·2	23·6	10·2	100·0
	Female	4·1	72·0	18·0	5·9	100·0
Manchester	Male	0·7	42·3	48·0	9·0	100·0
	Female	0·8	51·4	44·0	3·8	100·0
Liverpool	Male	2·4	56·0	29·6	12·0	100·0
	Female	3·0	60·5	28·2	8·3	100·0
Newcastle	Male	1·7	42·4	37·7	18·2	100·0
	Female	2·3	43·9	36·5	17·3	100·0
Glasgow	Male	2·0	65·9	23·7	8·4	100·0
	Female	2·2	73·4	19·7	4·7	100·0

Source: Census 1961, England and Wales: Workplace Tables, Tables 1 and 3B (General Register Office, London, 1966). Census 1961, Scotland: vol. 6, Occupation Industry and Workplace, Pt. III; Workplace, Tables 1 and 3B (General Register Office, Edinburgh, 1966).

work than men because they would tend to choose a workplace which is close to their residence, rather than a residence which is close to their work. Table 9.3 suggests that this cannot be the whole reason. It shows that, in each conurbation, the proportion of the total female population aged over 15 which is economically active (i.e. either in work or seeking work) tends to decline with distance from the conurbation centre, thus confirming another of the predictions made in the first section of the chapter. This suggests that the reasons for the shorter journey to work of females is that the homes of females in paid work tend to be located near their work, rather than work being found near the home.†

† Although, of course, the variation in economic activity may be due to the fact that it is much easier to find jobs near to home in the central city than in the suburbs.

TABLE 9.3

Proportion of the Total Female Population over 15 which is economically
active in each Subdivision of the Conurbation,
Major British Conurbations, 1961

Central city of the conurbation	Subdivision of conurbation		
	Conurbation centre (%)	Rest of central city (%)	Rest of conurbation (%)
London	59·0	49·7	42·3
Birmingham	62·3	46·7	42·5
Manchester	64·7	47·4	45·6
Liverpool	55·1	42·5	36·6
Newcastle	50·0	37·8	33·8
Glasgow	47·8	40·7	36·0

Source: Census 1961, England and Wales: Occupation, Industry and
Socioeconomic Groups, County Tables, Table 1 (General Register Office,
London, 1965, and 1966). Census 1961, England and Wales: Occupation
Tables, Table 26 (General Register Office, London, 1966). Census 1961,
Scotland: Occupation and Industry, County Tables, Glasgow and Lanark,
Table 1 (General Register Office, Edinburgh, 1965). Census 1961, Scotland:
Occupation, Industry and Workplace Tables, Part I; Occupation Tables,
Table 26 (General Register Office, Edinburgh, 1966).

TABLE 9.4

Proportion of the Total Population Resident in each Subdivision of the
Conurbation which is under 15, Major British Conurbations, 1961

Central city of the conurbation	Subdivision of conurbation		
	Conurbation centre (%)	Rest of central city (%)	Rest of conurbation (%)
London	13·9	20·1	20·6
Birmingham	21·2	23·5	23·3
Manchester	8·6	24·3	22·9
Liverpool	22·0	26·2	25·5
Newcastle	14·3	23·5	25·5
Glasgow	20·6	25·8	27·6

Source: See Table 9.3.

A third prediction of the theory is that the proportion of the population which is below working age will increase with distance from the city centre. The data in Table 9.4 only partly confirm this prediction.

The evidence relating to the fourth prediction, that household size will probably increase with distance from the city centre, will be examined in the final section of the chapter.

OTHER CITIES: FACTORIAL ECOLOGY AND
SOCIAL AREA ANALYSIS

Factorial ecology

According to the theory, there should be an ordered relationship between distance from the city centre and between certain family characteristics. The data in Table 9.5 shows, for the city of Melbourne, the way in which these family characteristics are each correlated with distance from the centre. Furthermore, it is obvious that these family characteristics are strongly related to each other, e.g. the proportion of women in the labour force will be negatively correlated with household size

TABLE 9.5
Zonal Patterns in Melbourne's Residential Structure, 1961.

	Zone			
	1	*2*	*3*	*4*
Percentage of the total population under 15 years of age	18·7	20·7	29·8	37·9
Percentage of the total population not employed	45·1	51·2	59·3	60·2
Percentage of persons aged 15 and over not married	35·9	30·1	22·6	16·9
Percentage of married women giving home duties as their occupation	47·8	56·7	66·2	64·7
Percentage of women aged 15 and over in the work force	47·7	40·1	31·0	31·0
Children aged 0–4 per 100 females aged 15–44	35·2	35·8	46·5	64·5

Source: Johnston (1969), Table IX. The zones are concentric and zone 1 is nearest the centre.

and with the proportion of the population under 15. These separate family characteristics might therefore be regarded as manifestations of a measurable variable denoting 'family status' or the 'stage in the life cycle of the family'. It is not, therefore, surprising that in the many recent studies of the geography of urban areas by factor analysis (and called studies of the factorial ecology of cities), it has usually been found that variables denoting these family characteristics load highly on to one of the factors accounting for much of the variance between the different subdivisions of the city.†

In a study of the factorial ecology of Toronto in 1951 and 1961, Murdie (1969) found that the following variables denoting family characteristics loaded highly on to a 'family status' factor in both years.‡

| | Factor loading | |
	1951	1961
Percentage of the population aged under 15 years of age	0·601	0·767
Percentage of females 14 years of age and over in the labour force	−0·771	−0·837

Other variables which loaded highly on to this factor were distance from the centre itself, and those variables which measure population density and which we would expect to be also correlated with distance from the centre from the analysis in Chapters 4 and 5, i.e.

| | Factor loading | |
	1951	1961
Distance from the peak land value intersection	0·596	0·603
Population density (natural logarithm)	−0·448	−0·478
Population potential	−0·506	−0·598
Percentage of dwellings:		
single detached	0·696	0·641
apartments	−0·804	−0·824

In several other studies of the ecology of cities by factor analysis, a 'family status' factor has been picked out and

† For a useful explanation of the techniques of factor analysis, see Rummel (1967).

‡ Murdie (1969). For convenience in interpretation I have reversed the signs on all the factor loadings.

shown to be distributed concentrically, i.e. the family characteristics of households have tended to vary with distance from the
centre in the manner predicted. The cities studied were
Chicago (Rees, 1970), Copenhagen (Pederson, 1967, reported
by Murdie, 1969), and Boston (Sweetser, 1961, 1962, reported
by Murdie, 1969).

Several studies of the factorial ecology of British cities have
been carried out, and a factor indicative of 'family status' or
'stage in the life cycle' has usually been picked out, but the
spatial distribution of the factors has not been analysed so that
the results cannot be shown either to confirm or reject the
theory. The cities studied were Cardiff and Swansea (Herbert
1970), Sunderland (Robson, 1969), and Liverpool (Gittus,
1965).†

Social area analysis

Factor analysis was first used in investigating the ecology of
U.S. cities in order to test the validity of the method of classi-
fication called social area analysis. 'Social area analysis was
developed by Eshref Skevky and a group of sociological colleagues as a technique for classifying census tracts according
to three indexes: economic status, family status, and ethnic
status'.‡

In social area analysis the family status index is constructed
from measures of fertility, women at work, and single-family
dwelling units. It can be seen that according to our theoretical
analysis these three measures would be correlated with each
other and with distance from the city centre. Hence, the results
of studies of cities using the techniques of social area analysis
are evidence which can be used to test the predictions of our
theory of residential location provided the spatial distribution
of the family status index is also analysed. There are two such

† One difficulty with studies of the factorial ecology of British cities is
that, as Herbert points out, the extent of public intervention in the housing
market is much greater in Britain than the United States, with the result
that the first component or factor extracted is usually associated with
public ownership of housing, and this factor often accounts for a large
proportion of the variance. The other factors denoting 'family status',
etc. therefore account for a smaller proportion of the total variance in
British cities than in U.S. cities.

‡ Murdie (1969, p. 17). See also Shevky and Bell (1955).

studies: Anderson and Egeland (1961) showed that 'family status' was distributed concentrically in the cities of Akron, Dayton, Indianapolis, and Syracuse N.Y., and McElrath (1962) showed that it was distributed concentrically in Rome (Italy). In Britain, Herbert (1967) carried out a study of Newcastle under Lyme by means of social area analysis, but did not analyse the spatial distribution of the indices.

VARIATION IN HOUSEHOLD SIZE

In the first section of the chapter we showed that, *ceteris paribus*, average household size may increase with distance from the city centre. The increase would be reduced and may be completely eliminated by the decrease in the number of workers per household with distance. Thus, even if it were found that household size decreased with distance from the centre, this would not necessarily refute the theory. On the other hand, the increase in household size is not really consistent with any other theory; in particular, it is inconsistent with the filtering-down theory which Burgess used to explain the existence of concentric zones around the central business district (CBD). Suppose that it were, in fact, true that the pattern of concentric zones is due to the lowest-income households living in the oldest property near the CBD, and the wealthiest living in the newest property on the edge of the city. If household size is not (positively) correlated with income (and there is no evidence to suggest that it is in Western cities),† it should not increase with distance from the centre. Indeed, it could be argued that it should decrease, since large families should locate in the inner areas of the city where, according to the filtering-down theory, the housing would be older and therefore cheaper.

We shall show that household size generally increases with distance from the city centre in British and American cities. This finding refutes the filtering-down theory as the sole explanation of the pattern of residential location in cities.

With respect to American cities, Muth (1969) regressed the

---

† The factor analytic studies which pick out two separate factors for socioeconomic status and family characteristics are positive evidence that the two are uncorrelated.

log of the average number of persons per household on the log of distance from the CBD. He used 1950 census data for a random sample of twenty-five census tracts in each of sixteen cities. The results are reproduced in Table 9.6. In 12 of the 16 cities, log of household size increases with log of distance from the centre, and in 7 out of the 12, the rate of increase is significantly different from zero at the 0·10 level. In the four cities where household size apparently declines with distance, the negative regression coefficient is not significantly different from zero. The results therefore reveal 'a distinct tendency for the average size of household to increase with distance from the CBD' (p. 189).

Table 9.7 shows the results of similar regressions for the largest British cities. Average number of persons per private household was regressed on distance from the CBD for seven large British cities, using 1961 census data for all wards in the central city (or all local-authority areas in the County of

TABLE 9.6

Results of Regressions. (Average Number of persons per household as a Function of Distance from the CBD, sixteen U.S. cities, 1950.)

City	Log linear regression coefficient	R^2
Flint, Mich.	0·070*	0·45
Nashville, Tenn.	0·057*	0·17
Omaha, Nebr.	0·034*	0·22
Louisville, Ky.	0·030	0·09
Cincinnati, Ohio.	0·028*	0·24
Dallas, Texas	0·019*	0·12
Fort Worth, Texas	0·018*	0·16
Houston, Texas	0·015	0·07
Philadelphia, Pa.	0·014*	0·12
Birmingham, Ala.	0·008	0·02
St. Louis, Mo.	0·006	0·01
Atlanta, Ga.	0·003	0·00
Baltimore, Md.	−0·002	0·00
Buffalo, N.Y.	−0·009	0·00
Miami, Florida	−0·010	0·01
Cleveland, Ohio	−0·017	0·10

Source: Muth (1969) p. 189.
* Significantly different from zero at the 0·10 level.

London). The results of linear and log linear regressions are shown.

The log linear regressions give results which are similar to Muth's. In 5 of the 7 cities, log of household size increases with log of distance, and in 3 out of the 5, the rate of increase is significantly different from zero at the 0·05 level. The two negative regression coefficients are not significantly different from zero. The results of the linear regressions are even more favourable. In all the cities, household size increases with distance, and in five cities, the rate of increase is significantly different from zero.

From these results we can conclude that household size does tend to increase with distance in British and American cities, and the results for British cities suggest that the relationship is linear rather than non-linear. Therefore the filtering-down theory can be rejected as a complete explanation of patterns of residential location in cities.

The low values of the coefficients of determination (R^2) in Tables 9.6 and 9.7 show that only a small proportion of the variation in household size is explained by variation in distance. In the case of Liverpool and Birmingham, distance from the CBD explains less than 1 per cent of the variation in household size. In Chapter 14 we shall show that variation in household size can be explained more fully if we take account of variations

TABLE 9.7

Results of Regressions. (Average number of persons per private household as a function of distance from the city centre, seven British cities, 1961).

City	Linear regression coefficient	R^2	Log linear regression coefficient	R^2
Edinburgh	0·318*	0·55	0·147*	0·47
Newcastle	0·299*	0·54	0·111*	0·42
Glasgow	0·184*	0·39	0·036	0·10
London	0·102*	0·33	0·119*	0·34
Manchester	0·079*	0·24	0·027	0·04
Liverpool	0·015	0·00	−0·034	0·05
Birmingham	0·005	0·00	−0·017	0·03

Source: Census 1961, England and Wales: County Reports, Table 3 (General Register Office, London 1963–64).
* Significantly different from zero at the 0·05 level.

in the social class of an area and in the amount of local-authority housing there, as well as the interaction with distance of these variables. But we must first understand the pattern of location of households in the city when there are numerous workplaces instead of just one. This is considered in Chapters 12 and 13, while some necessary preliminary work on variations in wages is covered in Chapter 11.

In Chapter 10 the theoretical analysis which has now been developed for the single-centre city is used to provide a possible explanation for various changes which have occurred in Western cities over time.

10 The Effects of Changes in Transport Technology: A Comparative Static Analysis

The methodological stance adopted in this chapter is rather different from that in the rest of the book. In the other chapters we use the empirical evidence to test the predictions of the theory. In this chapter we use the theory to explain various changes which are known, or believed to have occurred in cities generally.

Empirical evidence shows that in many cities the density gradients have become less steep over time. This is probably true of most Western cities, for central densities have scarcely increased over the years while the built-up area of most cities has increased enormously. There is also some evidence to show that the land-value gradient has become less steep over time in some cities, though it is more difficult to tell whether this is generally true.

Similarly, there is some evidence that the geographical distribution of incomes in cities has changed, and is changing, with the wealthier moving to the suburbs, leaving the central city to the poor. It is also believed – though with even less evidence – that the central business district (CBD) of the modern Western city is surrounded by a decayed inner ring of property which houses people at abnormally high densities, and which private enterprise does not find profitable to demolish and redevelop even though this property appears to have reached the end of its natural life. Yet, further out from the centre, relatively new property is demolished and re-developed at higher densities. We shall show that these things can be explained, in terms of the theory, by changes in transport technology.

THE CHANGING SPATIAL DISTRIBUTION OF INCOMES IN THE CITY

In Chapter 8 we showed that those households whose elasticity of demand for space with respect to the rate of pay is greater than a certain function of the rate of pay would tend to move out from the city centre with increases in income, while all those with elasticities less than this function would move in. This function we denote as $E(r)$, where:

$$E(r) = \frac{r \cdot v_t}{c_t + r \cdot v_t} = \frac{r \cdot v_k \cdot k_t}{c_k \cdot k_t + r \cdot v_k \cdot k_t} \qquad (10.1)$$

After an increase in the rate of pay, all those whose income elasticity is equal to $E(r)$ will remain at the same distance from the centre as they were located before. The graph of $E(r)$ is shown by the line OA_1 in Fig. 10.1. The spatial distribution of incomes in the city is determined by the distribution of incomes and income elasticities of demand for space relative to the points on this line.

Suppose that the urban system is initially in equilibrium with every household at its optimal location. What happens if travel speeds increase, e.g. through the electrification of rail-

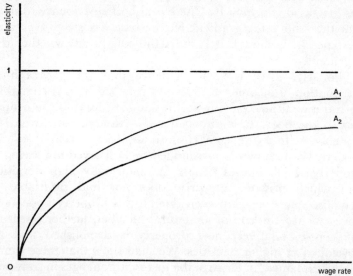

Fig. 10.1   The effect on the function E of an increase in the speed of travel.

ways, or the replacement of all-purpose roads by motorways or expressways? The consumer's valuation of time per hour $(r.v_t)$ remains the same, but the direct, financial, cost of travel *per hour* increases, even though the direct, financial, cost of travel *per mile* remains constant. More miles are now travelled in a given time. If $c_t(=c_k.k_t)$ increases but $r.v_t(=r.v_k.k_t)$ remains constant, the denominator of the fraction on the right-hand side of (10.1) increases while the numerator remains the same. Thus, at any given rate of pay, the elasticity at which the household's location remains the same, after a small change in the rate of pay, is lower than before the increase in transport speeds. The graph of the new function is shown by the line OA_2 in Fig. 10.1. The effect of the increase in travel speeds can be easily seen. Those households represented by combinations of rates of pay and income elasticities, denoted by points between OA_1 and OA_2 in Fig. 10.1, now have to reorder themselves with respect to distance from the centre: instead of those with higher incomes being located closer to the centre than those with lower incomes are, those with higher incomes now locate further from the centre than do those with lower incomes. In addition, those with combinations of income elasticities and rates of pay, denoted by points below OA_2 in Fig. 10.1, find that the attraction of a location near the city centre is weakened. The result of the increase in travel speeds is a reordering of the household location patterns in the city with the average high-income household being located further out relative to the average low-income household than before.

Other changes may also alter the spatial distribution of incomes. An increase in financial travel costs per mile, c_k, will have the same effect on the relative location of households as an increase in travel speeds. The increase in c_k causes c_t to increase. If c_t increases but $r.v_t$ remains the same, the value of $E(r)$ at any rate of pay falls, and we get the same reordering of high and low-income households relative to each other as described above.†

† Note that here we are only concerned with the location of households relative to each other. In the next section we shall be concerned with the spatial extent of the city and shall show that, while an increase in travel speeds leads to an increase in the area of the city, an increase in travel costs leads to a decrease.

An increase in the comfort of travel will also have the same effect on the spatial distribution of incomes. If the comfort o the travel activity increases, the consumer's imputed valuation of the time spent in that activity falls: time spent in travelling is resented less. By definition, therefore, v_t falls. By dividing numerators and denominators in (10.1) by v_t, the equation can be written in the form:

$$E(r) = \frac{r}{(c_t/v_t) + r}. \qquad (10.2)$$

It can be seen that a fall in v_t has exactly the same effect on the spatial distribution of incomes as an increase in c_t does, namely, a reordering of the population, with the wealthier locating relatively further out than they were before.

The only other change which might bring about an alteration in the spatial distribution of incomes is an increase in the real incomes of the population. If the cost of travel remains the same, the result will be movement inward of the high-income groups relative to the others. Suppose that in Fig. 10.1 the incomes of the population increase; some households with combinations of incomes and income elasticities which were formerly denoted by points above OA_1 will now be denoted by points below OA_1. Instead of ordering themselves with the highest-income households furthest from the centre, they will order themselves with the highest-income households nearest the centre. This can be seen more easily, perhaps, if we regard an all-round increase in real incomes as equivalent to a reduction in the real direct cost of transport relative to the value of time. The result must be increased centralisation of high-income groups, for this is nearly the opposite case to that discussed above where it was shown that an increase in the direct cost of transport leads to increased decentralisation of high-income groups.

It is probable that the dominant forces determining the redistribution of income groups in the city have been increases in transport speeds and improvement in comfort. The financial cost of travel per mile has probably fallen slightly over time, relative to incomes, but I would think that changes in transport technology have mainly led to improvements in speed and comfort rather than reductions in cost.

The above analysis provides a possible explanation (other than changing tastes) for the 'flight to the suburbs' which is said to be occurring in U.S. cities. It can be attributed to the large-scale construction of urban motorways which increases commuting speeds and comfort for the commuter. The analysis also provides an answer to the question: 'How do we attract the higher-income groups back to the central city?' The simple answer is, and perhaps it should not be taken too seriously: 'Make the transport system slow, cheap, and uncomfortable, and not fast, comfortable, and expensive.' For the lower-income groups, time is cheap relative to money; for the higher-income groups, money is cheap in comparison to time. With a fast, expensive, comfortable transport system the higher-income groups are willing to make long commuting journeys to the outskirts of the city because the time cost is low. The lower-income groups are unwilling to do so because the money cost is high. The poor therefore live near the centre, and the rich near the edge. With a slow, cheap, uncomfortable transport system the pattern is reversed. The higher-income groups are unwilling to travel to the periphery because the time cost is high and try to locate near the centre, the poor living at the edge.

The analysis explains not only the recent changes in location patterns in U.S. cities, but also the differences in patterns of location between the pre-industrial city and the industrial city, for the latter is likely to have a faster, more comfortable transport system. In the pre-industrial city 'the elite typically has resided in or near the center, with the lower class and outcaste groups fanning out toward the periphery' (Sjoberg, 1965, p. 216), while, 'in marked contrast . . ., the upper and middle socio-economic groups in the industrial city tend to reside beyond the city's core, leaving the central area to various low status groups, and elements of the elite as well' (p. 229f.). Schnore (1965) states that the traditional 'pre-industrial' pattern of location in Latin American cities seems to have begun to break down around the turn of the century, and to approach the pattern of the North American 'industrial city'.

It may be noted that in a primitive society, if all transport is by foot, the direct financial cost of travel per hour, c_t, is 0, and $E(r) = 1$. If, therefore, the income elasticity of demand for space is generally less than 1, the 'pre-industrial' pattern of location will necessarily emerge.

The analysis can help to explain a further phenomenon. Hoyt (1939) found that in U.S. cities the high-income sectors were frequently located on the fastest radial routes of the city. But from the above analysis we would expect that the ordering of households by income and distance from the centre would be different in sectors on fast transport routes to the ordering in sectors on slow transport routes. We would expect to find what Hoyt observed – a greater tenndecy for high-income households to locate in the sectors with fast transport routes.

CHANGES IN THE RENT, DENSITY, AND LAND-VALUE GRADIENTS

For any household the slope of the lowest attainable bid-price curve p_k^*, when the bid-price is a function of distance, is given by

$$p_k^* = -\frac{c_k + r.v_k}{q}. \qquad (10.3)$$

Let the household's optimal location be the point (p_1, k_1) where p_1 is the rent per space unit payable k_1 miles from the city centre; then the slope of the household's bid-price curve at this point is equal to the slope of the rent gradient and is given by (10.3).

Suppose that there is an increase in the speed of travel, k_t, e.g. through the electrification of railways used by commuters. As we showed in the previous section, if k_t increases and the comfort of the travel activity stays the same so that v_t is constant, then v_k decreases. That is, the value of time *per mile* decreases when the speed of travel increases, since each mile is travelled in a shorter time. After the increase in travel speed the household will have a new set of bid-price curves. The new bid-price curve passing through the point (p_1, k_1) will have a lower slope than the old curve, since the numerator will be less. Furthermore, the improvement in the consumer's welfare resulting from the time savings may lead to an increase in q, the quantity of space demanded at the price p_1, and hence, because of the increase in the denominator, to a further decrease in the slope of the bid-price curve at the point $(p_1, k_1.)$

Since the new bid-price curve cuts both the old bid-price curve, and hence the rent gradient, from below, the point (p_1, k_1) no longer indicates the household's optimal location. Its new location must be further from the centre of the city, and if the rent gradient remained the same, the household would move to this new location further from the centre. It can easily be seen that this argument holds for all households in the city. All households will find, after the increase in travel speed, that their old location is no longer optimal and that they could improve their welfare by moving further out. As a result, the demand for space in the inner areas of the city will fall and the demand for space further out will increase. Thus, the price of space in the inner areas will fall and the price of space in the outer areas will increase, and a new rent gradient with a less-steep slope will replace the old. In Fig. 10.2 curve AA denotes the old rent gradient, and BB denotes the new.

An increase in the comfort of travel or a reduction in the cost of travel will have the same qualitative effect on the rent

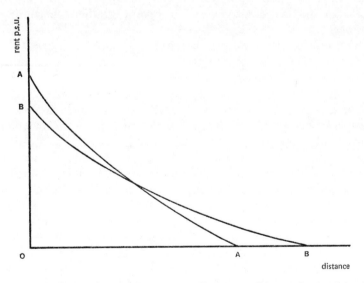

Fig. 10.2 The change in the rent gradient caused by an increase in the speed of travel.

gradient. In the first case the consumer's valuation of time, $r.v_t$, will fall, and if k_t and r stay constant this means that v_k must fall. In the second case, by definition, c_k falls. If c_k or v_k falls, the numerator of the fraction denoting the slope of the bid-price curve at the point (p_1, k_1) falls; moreover, because of the increase in the consumer's welfare, the denominator of the fraction – the quantity of space, q, may increase. Thus, the slope of the bid-price curve falls, households tend to move out, and the slope of the rent gradient decreases.

Note that, whereas a reduction in the cost of travel has here the same qualitative effect on the rent gradient as an increase in travel speed or an increase in comfort, in the previous section a reduction in the cost of travel was shown to have an effect on the spatial distribution of incomes opposite to the effects of increased travel speed or increased comfort.

The effect of a general increase in real incomes is difficult to predict. If the income elasticity of demand for space is generally greater than $E(r)$, the increase in incomes will cause the slopes of the bid-price curves to become less steep, and the rent gradient to become less steep. If the income elasticity is generally less than $E(r)$, the increase in incomes will lead to both the bid-price curves and the rent gradient becoming steeper.

The effects of an exogenous change – such as increase in the speed of travel, as discussed above – can be shown most clearly, if less generally than above, if it is assumed that all the households in the city have sets of bid-price curves which, when drawn, appear identical. It is not assumed that each bid-price curve represents the same utility level for each household; only that if each household were asked to outline the bid-price curve associated with a given rent level at the city centre (or some other location), each household's bid-price curve would be identical to that of any other household.

Suppose that some of the set of bid-price curves, as they would be drawn before the increase in travel speed, are shown in Fig. 10.3 by the lines XW, BZ, and AA'. The old rent gradient is shown by the line AA' and, as shown in Chapter 6, lies wholly along a bid-price curve. If there is an increase in travel speed, the old set of bid-price curves is replaced by a new set. Three new bid-price curves are shown by the lines XA', BB', and AY.

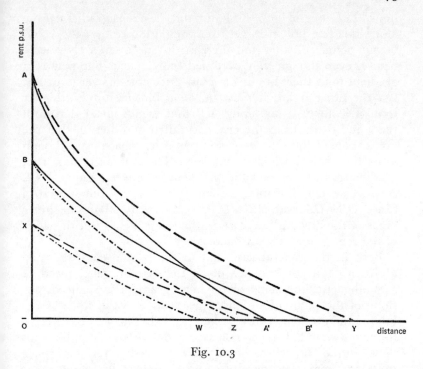

Fig. 10.3

These three curves denote equivalent utility levels to the old curves, XW, BZ, and AA', respectively. For example, curve XW implies that, before the speed increase, each household would be indifferent between a rent OX at the city centre and any combination of rent and distance from the centre represented by the points on the line XW. The slope of the curve reflects the rate of increase in the cost of living at a distance from the city centre. After the speed increase the cost of living at any given distance is less, and the rate of increase in the cost falls. Hence, the slope of the new bid-price curve is less than the slope of the old, and the new curve XA' shows that, after the speed increase, each household would now be indifferent between a rent OX at the city centre and any combination of rent and distance from the centre represented by points on the line XA'.

After the speed increase the bid-price curve which represents the utility level equivalent to the old rent gradient AA' is the line AY. It might be possible for the new rent gradient to lie along this line but, if it did, the population of the city would be no better off after the increase in speed than they were before, even though they occupied more space. The new rent gradient must therefore lie on a bid-price curve lower than AY. On the other hand, it must lie on a bid-price curve higher than XA', for if it lay along XA' that would mean that both rents and densities in the city had fallen and, hence, that the households of the city were occupying less space even though rents had fallen, which is unlikely. Therefore, the new rent gradient must lie along a bid-price curve between XA' and AY. A possible equilibrium gradient is shown by the line BB'. Since $OB < OA$ and $OB' > OA'$, rents must fall in the inner areas of the city and increase in the outer areas, and the slope of the rent gradient must decrease.

Both in the general case and in the special case discussed above if, as a consequence of an increase in travel speed or anything else, the slope of the rent gradient is reduced, the slope of the population-density gradient and the slope of the land-value gradient will also be reduced. The graphical technique shown in Fig. 4.4 can be used to derive a new equilibrium-density gradient, where density is expressed in terms of space units per acre. The density of space units falls in the inner areas and increases in the outer areas. As a result of these changes the density of population will fall in the inner areas of the city and increase in the outer areas, and a new population-density gradient will result which will be less steep than the old gradient, the change being similar to that in the rent gradient shown in Fig. 10.2.

The graphical technique of Fig. 4.5 can be used to show that when rents fall in the inner areas of the city, so land values will fall, and when rents increase in the outer areas of the city, so land values will increase. Hence, the increase in travel speed will result in a new land-value gradient which will be less steep than the old.

It should be noted that even though rent gradients may become less steep, the land-value gradient and density gradient may alter in a different way if changes in building technology are occurring at the same time. A change in building technology which allows high densities to be built more cheaply, while leaving the cost of building at low densities the same, would cause the density and land-value gradients to become more steep, even though there would be no change in the rent gradient. Also, rents may fall very little, if at all, in the centre of the city if the population of the city increases at the same time as the other changes are occurring. The increase in the population would cause rents to be bid up over the whole area of the city so that, although the slope of the rent gradient would decrease, rents at the centre would scarcely fall.

In most cities for which empirical evidence is available, the gross population-density gradient has generally declined in slope over the last century or so. Evidence for Chicago shows that the land-value gradient has been declining in slope there since the nineteenth century (see Mills (1969) and Yeates (1965)). The dominant forces determining changes in the rent, density, and land-value gradients would seem to be increases in the speed and comfort of travel, just as these increases have caused changes in the geographical distribution of incomes.

Clark (1951) graphed five gross-density gradients for London for dates between 1801 and 1841. These graphs are reproduced here as Fig. 10.4. Between 1801 and 1841 no change in the slope of the curve is observable, only an upward shift indicating overall population growth; between 1841 and 1871 a slight reduction in the slope can be seen but the major decline in the slope occurs between 1871 and 1921, and this is probably due to the construction of the underground railway system and the beginnings of motorised transport. The reduction in slope continues between 1921 and 1941.

Data for Chicago prepared by Rees (1970a) show that there the reduction in the slope of the gross-density gradient was greatest in the decades 1910 to 1920 and 1940 to 1950, the first reduction being presumably due to improvements in mass transit, and the second a result of the beginnings of mass automobile ownership. Rees has also calculated density gradients for several cities over various periods of time (see Berry and Horton (1970, Fig. 9.8)). With the exception of Calcutta the gradients in all the cities appear to decline over time, though the population density in the centre may increase or decrease. I say 'the gradients appear to decline' because in

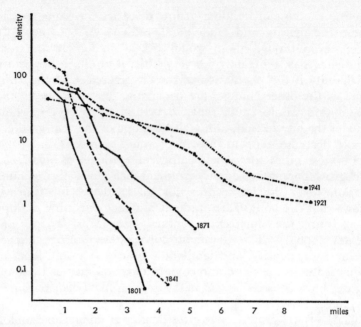

Fig. 10.4   Density gradients – London, 1801 to 1941. *Source:* Clark (1951).

common with most researchers Rees assumes that the slope of
the density gradient is denoted by the coefficient of distance in
the exponent of a negative exponential density function. As noted
in Chapter 5, a negative exponential function relating density
to distance can be fitted to the data for any city. This function
has the form:

$$D(k) = d.e^{-bk}$$

where $D(k)$ is population density at distance k from the city
centre, and d and b are constants. Since $e^{-bk} = 1$ when $k = 0$
d denotes the estimated density at the centre of the city. In its
logarithmic form, this equation is fitted to the data:

$$\log D(k) = \log d - b.k.$$

The rate of change of $\log D(k)$ with distance is equal to $-b$;
most researchers call this the 'density gradient'. But $-b$ is
not the rate of change of density with distance. This is given by:

$$D'(k) = D_k = -b.d.e^{-bk} = -b.D(k).$$

If both b and d are functions of time, D_k will be a function of time, and the rate of change of D_k over time is given by:

$$D_{kt} = [b_t.d \ (b.k - 1) - b.d_t] \ _e^{-bk}.$$

Only if d increases over time while b stays constant will the true density gradient necessarily become steeper everywhere. If d increases but b decreases, the true gradient, denoted by D_k, may become less steep near the centre and more steep further out.

Failure to make explicit the distinction between b and D_k can lead the researcher to make misleading statements. Thus, Muth (1969) and others have found that 'the density gradient', b, falls with city size. But this does not necessarily mean that the true density gradient becomes less steep, for the extrapolated central density, d, increases with city size. Even if b falls, the increase in d may mean that the true density gradient becomes more steep. Mills (1970, Table 5) estimates that an increase of one million in the population of a city in the United States leads to a reduction in b of 0·00445 and an increase in d of 671. The difference in the order of magnitude of the two changes suggests that the effect of the increase in d will be dominant, and the true density gradient will become more steep rather than less steep.

Similarly, Mills (1970) also finds that b appears to increase rather than decrease over time, when changes in population and income have been allowed for (though it should be noted that the coefficient of time in his equation is clearly not significant). Yet one of his two estimates shows that d may decrease over time. If d decreases and b increases, the slope of the true density gradient may, in fact, decrease rather than increase over time.

THE PROCESS OF ADAPTATION

The change from one equilibrium population-density gradient to another cannot take place overnight even though the increase in the speed or comfort of travel may be more or less instantaneous. The cost of redevelopment means that the change from one equilibrium to another will take place over a long period.

Suppose that an area of land near the centre of the city was developed at a density OD_1 (space units per acre) before an improvement in transport technology, when the rent per space unit at that location was equal to OP_1. Suppose now that the speed or comfort of travel increases and that the rent per space unit offered at that location drops to OP_2 as the rent gradient changes to adjust to the new condition. The density at which the site would be redeveloped (if it were redeveloped) would fall to OD_2. The situation is illustrated in Fig. 10.5 where SS is the developer's marginal-cost curve.

The developer has no incentive to redevelop the site, however. If he allows the existing development to remain, his income from the property will be $OP_2 \cdot OD_1$ per acre. If he redevelops the site his income will be $OP_2 \cdot OD_2$ per acre, a fall of $OP_2 \cdot D_2D_1$. Since redevelopment means a fall in income it is worth while for the developer to let the existing buildings remain on the site. This will be true for all the inner areas of the city in which rents fall.

The situation can also be illustrated graphically using one of the models developed in Chapter 7. In Fig. 10.6 the expected net income (excluding the cost of ground rents) from the development is shown by curve A_1A_1 and the ground-rent by the horizontal straight line G_1G_1. The expected life of the

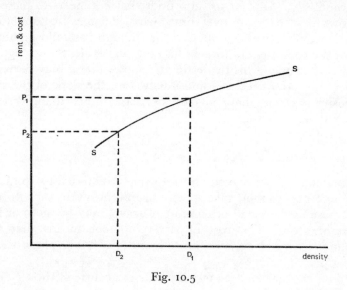

Fig. 10.5

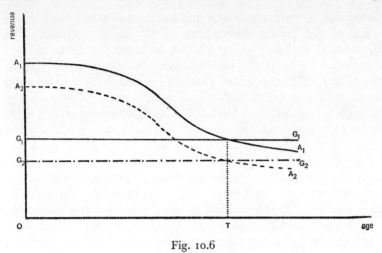

Fig. 10.6

building is therefore OT years. If, after construction of the building, transport speeds change and the rent per space unit which is payable at that site falls, the net income which the developer could obtain after redevelopment will also fall, e.g. to A_2A_2. Note, however, that the value of the site will also fall, and hence, so will the ground-rent.

Suppose that the new ground-rent is G_2G_2 so that the expected life of a new development on the site is OT years, the same as the original expected life of the old building. If the site is not redeveloped, so that the density is always higher than it would be if the site were redeveloped, then, because of the higher density, the developer's net income from the old building will always be higher than that indicated by curve A_2A_2 (even though, because of the fall in rents, the net income will always be lower than that indicated by curve A_1A_1). Hence, the net income from the property will not fall below the ground-rent until well after OT_1 years have elapsed. Thus, buildings in the inner areas of cities, for which rents have fallen because of increases in travel speeds, will tend to be left standing long after the buildings' 'normal' expected life.

This analysis thus provides an explanation for the existence of decayed inner rings in the inner areas of cities. This explanation is consistent with the difficulty of persuading private enterprise to redevelop these areas at a time when the government or the public thinks they should be redeveloped. The theoretical explanation states that it is profitable to private enterprise to keep the old buildings standing, even though their state of decay means that only a relatively low rent per space unit can be charged, because the site can be occupied at such a high density. Certainly this explanation would seem to be in accord with the existing image of housing in the inner ring. On the other hand, it should be noted that the condition of the housing may depend on the incomes of the occupiers, as suggested in Chapter 7; the condition, but not the density of property in the inner ring, could therefore be explained by the fact that it is the location of those with very low incomes.

The analysis also explains the tendency for the true density gradient to be more steep in older cities. Muth estimated negative exponential density gradients for forty-six U.S. cities, and it has been found that both b and d are positively correlated with the age of the city, where age is defined as the number of years since the city's population reached 50,000 (see Muth (1969, p. 154) and Berry and Horton (1970, p. 288)). If both b and d increase with age, then the density gradient must be steeper in the inner areas of older cities.

In the outer areas of the city the rate of adjustment to the changed situation will be much faster. Suppose that an area of land towards the edge of the city were developed at a density OD_2 (space units to the acre) when the rent per space unit at that location was equal to OP_2. If the equilibrium rent per space unit offered at that location increases to OP_1 the density at which it would be most profitable to redevelop the site would increase to OD_1 (see Fig. 10.5). If the developer allows the existing development to remain, his income from the property will be $OP_1 . OD_2$. If he redevelops the site, however, his income will be $OP_1 . OD_1$, an increase of $OP_1 . D_2D_1$. Since redevelopment means an increase in income, it will be worth while redeveloping the site to a higher density fairly quickly. This will be true of all the outer areas of the city in which rents increase. Buildings in these areas will often be demolished well before the end of their expected 'normal' life to make way for the new developments at higher densities.

SUMMARY

In this chapter we have used the theoretical analysis of residential location in a single workplace city to explain the changes in the urban structure which have occurred over time in terms of changes in transport technology, particularly increases in the speed and comfort of travel.

Using the method of comparative statics we have shown that reductions in the slope of the density, land value, and rent gradients, changes in the geographical distribution of incomes in the city, and the existence of a decayed inner ring or 'twilight zone' will all be necessary consequences of increases in the speed and comfort of travel.

This chapter concludes our analysis of location in a city with a single workplace. In the next chapter we investigate the variation in wage levels which can be expected to occur both between cities and within cities. This is a necessary foundation for the analysis – beginning in Chapter 12 – of patterns of residential location in cities with several workplaces.

11 On Wages

In the earlier chapters we investigated patterns of residential location by using the simplifying assumption that each city has one single central workplace. In later chapters we shall relax this assumption, and study patterns of location in cities where there are workplaces other than the central business district (CBD). As a necessary preliminary to this study we shall show in this chapter that, when workplaces are scattered over the city, wage rates should decrease with distance from the city centre. The empirical evidence on this is not conclusive, however.

THE WAGE GRADIENT

Both Moses (1962) and Muth (1969) have established by deductive reasoning that within cities wage levels should vary with distance from the city centre, and that in a city which has a single dominant centre there should be a wage gradient corresponding to the rent gradient, i.e. wages should decline with distance from the city centre.† In this section of the chapter we shall set out a version of the argument using the same tools of analysis as we use for the theory of residential location.

To simplify the argument we assume that the households in a city have sets of bid-price curves which are indistinguishable in shape. If there is initially a single central workplace the rent gradient will lie along a single bid-price curve. Suppose now that a new subcentral workplace is set up between the centre and the edge of the city. Some of those working

---

† Note that they derive a smooth, monotonic wage gradient from the smooth, monotonic rent gradient of a monocentric city. But wages will only fall monotonically with distance if the subcentral workplaces are so small that the demand for residential space by those working there does not distort the rent gradient. This is obviously unrealistic and, wages will therefore fall irregularly, with minor peaks at large subcentres.

at the centre can now work at the subcentre. In Fig. 11.1 we show a cross-section through the two sets of bid-price curves along the radius of the city through the subcentre.

AA' indicates the rent gradient, and $A'B$ indicates the rent per space unit on the plain surrounding the city. The subcentre is located OD miles from the city centre. The bid-price curve for those working at the subcentre, $C''CC'$, is identical in height and shape to the bid-price curve for those working at the centre along which the rent gradient lies (i.e. AA'). CC' and AA' are parallel and OA equals DC. Any household would be indifferent between work at the centre and a rent and residential location indicated by a point on the line AA' and work at the subcentre and a rent and residential location indicated by a point on the line $C''CC'$.

Suppose that the rent gradient remains unchanged after the new subcentral workplace is set up. It can be seen that, if workers at the centre and the subcentre are paid at the same rate, the workers at the subcentre will be better off since they

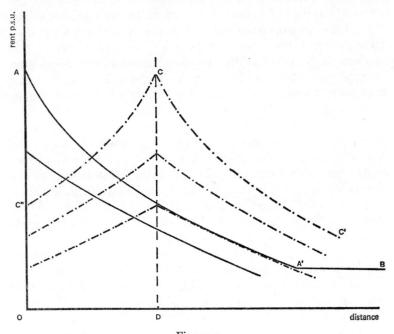

Fig. 11.1

can choose a residential location such that the rent gradient is tangential to a bid-price curve which is lower than $C''CC'$. The labour market cannot be in equilibrium if the same wage rates are paid at both workplaces and it will only be in equilibrium if a lower wage rate is paid at the subcentre. Furthermore, the argument can easily be extended to show that the further a subcentre is from the centre, the lower will be the equilibrium wage level at that centre. Hence, we can predict that wages will decline with distance from the city centre in large cities, although the decline will not necessarily be regular, or even uninterrupted. The wage levels in major subcentres may be almost as high as in the city centre.

Empirical evidence

The most favourable interpretation of the available evidence is that it is inconclusive. The British evidence, as we shall show later, demonstrates the existence of a wage gradient only in Greater London. In New York the wage gradient appears to be non-existent. Segal (1960) reports that in 1954 average annual earnings of production workers in the counties of the New York metropolitan region outside New York City were 18 per cent higher than the average earnings of production workers in New York City. In fifteen of the suburban counties average earnings were higher than in New York City, while they were lower in only two. The difference may, of course, have been caused by differences between the type of industry in New York City and the type of industry in the suburban counties. Industries which pay high wages might be located in the suburbs. The evidence relating to particular industries shows that this cannot be the complete explanation. In the case of the garment industries and six other industries, earnings were lower in the suburbs than in New York City, but for thirteen other industries for which data were available, hourly earnings were higher in the suburban counties.

Rees and Schultz (1970) found in their study of the labour market in Chicago that there was a 'strong positive association of wages with distance traveled to work'. However, they describe this as 'a compensating differential needed to draw enough workers to the less accessible establishments' (p. 219). There is no implication that there is an association between wages and distance from the city centre, rather the reverse, since accessibility is presumably inversely related to distance from the city centre. For blue-collar occupations they find that there is a wage gradient but it is one 'with its highest point at the southeast of the Chicago area, where the concentration of heavy industry is greatest, then sloping downward toward the northwest' (p. 220).

The evidence for the British conurbations, other than London, also fails to support the theoretical predictions. There is no significant difference between male clerical salaries in the central cities and in the other parts of the conurbations. In the *Clerical Salaries Analysis: 1962*, published by the Institute of Office Management, there were thirty-two grades for which the median salary in a central city could be compared with the median salary in the rest of the conurbation (seven in the West Midlands, nine in South-East Lancashire, five in Merseyside, seven in Clydeside, and four in Tyneside). In three cases there was no difference, in ten cases the median salary in the central city was the higher, and in nineteen cases it was the lower. The sign test described by Siegel (1956, p. 68) shows that the difference is not significant at the 5 per cent level.

For female clerical salaries there were thirty-six observations. In eleven cases the median salary was higher in the central city but in twenty-five cases it was higher in the rest of the conurbation. The difference is significant at the 5 per cent level but it is in the opposite direction to that predicted by the theory.

Only the evidence for London clearly confirms the theoretical prediction. In the *Clerical Salaries Analysis: 1962*, median salary levels are given for two parts of the central business district (City and West End) and for four suburban sectors, one of which, the south-east sector, includes a part of the CBD south of the Thames.† The variation in salary levels for male and female clerks in Greater London is shown in Figs. 11.2 and 11.3. Salary levels in the City and the West End are clearly higher than in the suburbs, and the salary level in the south-east sector, partly in the CBD, is intermediate between that in the CBD and that in the other suburbs.

† In later issues of the *Clerical Salaries Analysis* the south-east sector is split into an inner and an outer part.

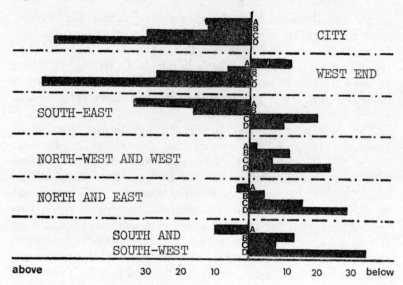

Fig. 11.2. Variation in wage levels of male clerks in Greater London 1962. Shillings above and below London Average for four clerical grades. *Source:* Institute of Office Management (1962, p. 22).

Thus, except in the single case of London, there is no evidence of any intra-city wage gradient. Why should this be so? There are several possible explanations. Firstly there is both a tendency for earnings to be positively correlated with plant size,† and a tendency for plant size to increase with distance from the city centre.‡ Together, these two relationships may eliminate or reverse the slope of any negatively sloped wage gradient. It may be noted that, according to Evely and Little (1960), there has been a marked rise in plant size in Britain since 1939, so that the effect of differences in plant size would have become more significant since then.

A second possible explanation is that there are advantages in working near the centre which balance the housing and travel costs involved in working there. It may be possible, for example, that proximity to the shopping facilities of the city centre is regarded as a benefit for which it is worth paying

† See Ministry of Labour (1959), Lester (1967), and Mackay *et al.* (1971). Rees and Shultz (1970) do not find this in Chicago, however.

‡ Hoover and Vernon (1959), Martin (1969).

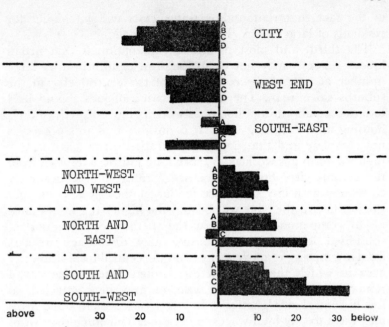

Fig. 11.3 Variation in wage levels of female clerks in Greater London, 1962. Shillings above and below London average for four clerical grades. *Source:* Institute of Office Management (1962, p. 23).

higher travel and housing costs. Against this, it may be noted that while central London's facilities are obviously better than those anywhere else, employers there have to pay wages which are considerably higher than elsewhere. Furthermore, this explanation is incompatible with the fact that wages increase with city size. One would expect that if the benefits of proximity to the facilities of the central city explain the general absence of intra-city wage gradients, the benefits of proximity to the facilities of large cities would result in the absence of any relationship between wage levels and city size. But there is ample evidence that wage rates do increase with city size, for Britain (Evans, 1972c) for Switzerland (Widmer, 1953) and especially for the United States (Mansfield, 1957; Fuchs, 1967; Mattila and Thompson, 1968; Hoch, 1972), and the relationship can be justified in terms of the theory of residential location in a monocentric city as a consequence of the fact that rent and/or travel costs will be higher for residents of larger cities (Evans, 1972b, 1972c).

The third, and most plausible, explanation is that urban labour markets may not be in long-run equilibrium. A large number of firms have moved out of the central city to the suburbs since 1945. The decentralisation of jobs should lead to a decentralisation of homes as workers move to reduce their housing and travelling costs. It is possible that this process is not complete and that newly decentralised firms have to pay high wages to get workers to make long journeys to work from the central city. Furthermore, wage rates will be positively correlated with city size, since the larger the city is, the longer the journey to the edge will be and the higher the wages must be to compensate. In Britain the method of allocation of subsidised local-authority housing may discourage workers from attempting to move away from the central local-authority area in which they live. In the United States members of minority ethnic groups may wish, or may be compelled, to live in central city 'ghettos'.† In either case decentralised firms may have to pay high wages to get the labour force they wish.

Note that, according to Segal (1960), 'as recently as 1939, average annual earnings of production workers in New York City were higher than such earnings in the other [metropolitan] counties taken as a group' (p. 140). This lends support to the view that the absence of a wage gradient in 1953 was due to the decentralisation of firms preceding the decentralisation of homes.

It is also likely that the plants which have caused the decentralisation of jobs are large new plants. If, then, large plants pay higher wages because they are less accessible, this explanation of the lack of a wage gradient is the same as the first, and we have instead an explanation of the positive correlation between plant size and wage rates. If, on the other hand, large plants pay higher wages for some other reason, for example because they have better selection procedures than small plants, and hence a higher-quality labour force, then the first and third explanations are alternative and complementary.

Finally, we should note Malamud's finding that 'offices

† Kain (1968) argues that the decentralisation of firms has considerably worsened the employment prospects of Negroes.

compensate their employees for longer travel times to work by reducing hours and leaving wages unchanged' (Malamud, 1971, p. 112). 'Hours of work at suburban offices are longer than in downtown offices' (p. 111). As a result, weekly wage rates may give no indication of a wage gradient even though it would be seen to present if hourly wage rates could be examined.

CONCLUSIONS

We have found that, while there is a well-documented, positive relationship between wage levels and city size, there is no clear relationship between wage levels and distance from the city centre in any city but London. The most plausible explanation for the absence of wage gradients is that suburban firms have to pay high wages to draw their labour force from the central city, from which, for whatever reason, households are slow to move. This explanation is compatible with the existence of a positive correlation between wage levels and city size, since the larger the city, the longer the average journey to work whether the journeys be from the edge to the centre or from the centre to the edge.

In the next two chapters we analyse the pattern of journeys to work in large cities in which there is more than one workplace. We assume that there is a wage gradient, both because it is difficult to analyse the pattern in disequilibrium, and because we shall use London for most of the empirical analysis, and as we have shown, in London a wage gradient exists.

12 The Journey to Work – I: Multiple Nuclei

The theory used in earlier chapters in the description and analysis of patterns of residential location was based on the simplifying assumption that each city had one single central workplace. But, as Harris and Ullman (1945) point out, 'in many cities the land-use pattern is built not around a single center but around several discrete nuclei'. In this chapter and the next the theory will be elaborated to take into account the existence of these 'multiple nuclei', and the predictions of the theory will be tested against empirical evidence.

Here, we still assume that the central business district (CBD) is the most important workplace in the city but we also assume that there are several important subcentres. In the first section we develop the theoretical argument and develop predictions about the pattern of the journey to work in the city. In the second section we compare these theoretical predictions with the evidence relating to the journey to work in secondary employment areas in London, and from residential areas in London. In the final section we show how, using the notion of entropy maximisation, the theory can be used to derive a stochastic model of the type used to simulate the distribution of person trips in transport planning.

THEORETICAL ANALYSIS

In earlier chapters rent gradients and bid-price curves have been used as the tools of analysis. Implicit in the analysis has been the assumption that rent was a function of distance from the CBD alone and that the analysis need only be conducted in terms of two dimensions. In the present context, however, rent must be thought of as a function of location, and the analysis must sometimes be conducted in terms of three dimensions. Instead of the rent gradient we have the rent surface, which in a city with no secondary employment centres would be a cone. Instead of the set of bid-price curves of a household we have a set of bid-price cones centred on the household's workplace.†

As a first approximation we assume that there is a major central workplace (the CBD) and a single minor subcentre. The number of workers at each centre is assumed to be fixed.‡ To simplify the exposition we will also assume that the workers at the CBD have identical sets of bid-price cones and that the workers at each subcentre have identical sets of bid-price cones (though the set of bid-price cones of a worker at the CBD is not necessarily identical to the set of bid-price cones of a worker at the subcentre).

A cross-section through the two sets of cones along the line of the CBD and the subcentre would reveal two sets of bid-price curves, as shown in Fig. 12.1. In equilibrium, the rent surface for those working at the CBD must lie along one of their bid-price cones, and the rent surface for those working at the

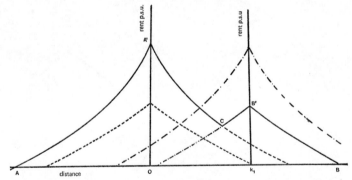

Fig. 12.1 A cross-section through two sets of bid-price cones.

† It is assumed that there is only one worker in the household. We ignore the case in which there are two or more working in different places, when the bid-price surfaces would no longer be cones (see first footnote, Chapter 13).

‡ Note that the location of the subcentre is not assumed to be determined by reference to the rent surface. As Parry Lewis has pointed out, Alonso appears to assume that the urban firm chooses a non-central location to maximise profits — taking the rent surface as given — but that the rent surface is determined by the location of residences about a single central workplace. (Medhurst and Lewis, 1969, p. 70 f.).

subcentre must lie along one of their bid-price cones. It follows that a possible cross-section of the rent surface along the line of the CBD and the subcentre is indicated by the line $AA'CB'B$ (for simplicity we assume that the rent on the plain surrounding the city is equal to zero). Along this line the residential location of CBD workers, and the associated rents, are denoted by the line $AA'C$, and the residential locations and rents of workers at the subcentre are denoted by the line $CB'B$. The cross-section of the rent surface must have this serrated appearance if subcentres exist. If it were a smooth surface there would be no location at which the subcentre workers would be able to outbid the CBD workers in their attempt to locate on the lowest bid-price curve. As it is, CBD workers can outbid subcentre workers for locations between A and C and a subcentre worker can outbid a CBD worker for a location between C and B. A CBD worker locating between C and B would not be on his lowest possible bid-price curve, while a

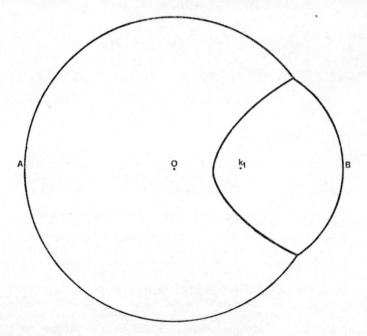

Fig. 12.2 Geographical distribution of the residences of workers at the CBD and a subcentre.

subcentre worker locating between A and C would not be on his lowest possible bid-price curve.

The actual bid-price cones which form the rent surface will be determined by the relative numbers of workers at each centre and the amount of space that they require for residences. In plan form, the allocation of the land area of the city between the two competing groups will be approximately as shown in Fig. 12.2. The spatial allocation of the residential locations of workers at the two centres results from the plane projection of the intersection of two cones. Those working in the subcentre will locate in a sector of the city lying along the radius through the subcentre. The shape of the sector will depend on the proportion of the city's workers who work in the subcentre. If a very small proportion work there, the land area they require may be located in a narrow sector as in Fig. 12.3, very few of the households being located between the subcentre and the CBD. If a very small number indeed work in the centre

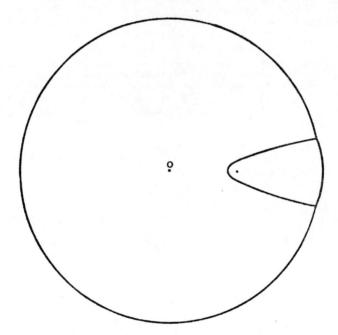

Fig. 12.3 Geographical distribution of the residences of workers at the CBD and a subcentre.

the rent surface for the city may be scarcely affected and may approximate to a smooth cone. The workers in the subcentre will then locate their residences along the radial line from the CBD through the subcentre between the subcentre and the edge of the city.

Some other results can be drawn from this method of theoretical analysis. Suppose that the workplace indicated in Figs. 12.1 and 12.2 were closer to the CBD than OK_1 miles but required the same working population. The sector in which the working population could live would be subtended by a smaller angle, and fewer workers would live between the workplace and the CBD. If it were adjacent to the CBD then obviously no workers would live between the CBD and the subcentre. If, on the other hand, the subcentre were moved towards the edge of the city, the area in which its workers live would be subtended by a larger angle and would approach a

TABLE 12.1

Mean distances to work by sectors of residence
and employment – London, 1962

	Mean distance to work (in miles)	
Traffic sector	Persons employed in sector	Persons resident in sector
Central area – North	6·34	1·37
Central area – South	5·53	1·62
L.C.C. area – North	2·90	2·27
L.C.C. area – South	2·65	3·23
Essex	2·61	4·45
Middlesex – West	2·96	3·84
Middlesex – East	2·42	3·87
Surrey	2·57	4·90
Kent	2·39	5·60
Hertfordshire	3·24	4·82

Source: London Traffic Survey (1964), vol. 1, Table 6.34. The traffic sectors are mapped in Fig. 3.1. and 3.2 of the same volume.
Note: In the above table the sectors are listed approximately in order of distance from the main centre of employment.

circle, with a much greater proportion of the workers living between the subcentre and the CBD. Thus, the further a place of work is from the CBD, the shorter will be the average journey to work. On the other hand, since in general those living near the centre will travel only to the CBD or nearby workplaces while those living further out will still travel to the CBD and to intermediate workplaces, we can also conclude that the further a residential area is from the CBD, the greater will be the length of the average journey to work. Evidence on the journey to work gathered in the course of the London Traffic Survey in 1962 and presented in Table 12.1 suggest that in London, at least, these two predictions are fulfilled.

EMPIRICAL EVIDENCE

There would seem to be two possible ways of confirming the theoretical predictions set out above. Firstly, the shape of the rent surface could be examined to discover whether the minor peaks in the surface were centred on the employment subcentres. Thus, Yeates (1967) shows that land values in Chicago are inversely related to distance from the CBD and other independent variables. The collection of information on rent surfaces is considerably more difficult and time consuming than the collection of data to estimate a rent gradient, however, and has only been done in a very few cities.†

Secondly, the pattern of the journey to work in employment subcentres could be investigated to discover whether the predicted sectoral pattern indeed occurs. In fact, this sectoral pattern has been noted on several occasions. Carroll (1952) after a review of the survey data then available concluded that, while 'the residential distribution of persons employed in central districts tends to approximate that of the entire urban area population . . ., residences of persons employed in off-center workplaces are concentrated most heavily in the immediate vicinity of the place of work' (p. 272). Again, in the Detroit metropolitan area transport study (1953) it was found in an investigation of travel to four non-central factories that each factory tended to draw its workers 'more heavily

† Knos (1968), for example, portrays the rent surface of Topeka, Kansas, but does not relate the minor peaks in the surface to employment centres.

from residential areas which are on the opposite side of the plant from the center of the city'. The most intensive study of the journey to work by non-central workers was carried out by Taaffe, Garner, and Yeates (1963) using data derived from the Chicago area transport study. They attempted to simulate the pattern of journeys to work to an employment subcentre in the western suburbs of the Chicago area located on the boundary of the City of Chicago. Their findings are consistent with the theory set out above, in that they found that, even when transport costs and alternative employment opportunities had been taken into account, the number of commuters to the workplace from the suburbs on the opposite side of it from the city centre 'is still greater than one might expect'.

A statistical test of the suggested hypotheses is provided by Boyce's 1965 study of 'The Effect of Direction and Length of Person Trips on Urban Travel Patterns', again using data obtained as part of the Chicago area transport study. A random sample of zones was drawn from each concentric ring of zones about the Chicago CBD. The number of trips between each zone was hypothesised to depend on the distance between the zones, the direction of the destination zone relative to the CBD and the origin zone, and it was also hypothesised that there might be some interaction between the two effects. Boyce found that 'a trip direction effect is significant for transit trips, higher in the direction of the center, but not for arterial trips. A significant interaction is also noted for transit trips indicating that the effect of trip length on mean volumes differs by direction of trips' (p. 77).

Thus, even though Boyce's data refers to all person trips and not just the journey to work, his findings serve to confirm the theoretical predictions. The absence of any trip direction effect for arterial trips is probably best explained by the fact that those choosing to travel towards the centre are best served by rail transit while travel in other directions is easier by road. As Boyce remarks, 'one can speculate that if rail transit service were avilable in all directions . . . ., the bias in trip direction toward the center would be reduced. At the same time, such a rail transit network might result in fewer auto trips in non-central directions resulting in a trip direction effect for auto trips' (p. 79).

For British cities, the earliest study of the journey to work was carried out by Liepmann (1947). Her concern was mainly with its cost, but the maps which she publishes of the residential locations of the employees at three large factories, two in London and one in Birmingham, demonstrate the existence of a pronounced sectoral pattern. Later studies of the journey to work in British cities have relied on Census data, e.g. the studies of the journey to work in Britain by Lawton (1968), in London by Westergaard (1957) and Wabe (1969), in Manchester by Green (1959), and in Lancashire by Warnes (1968). Of these, only Warnes was concerned with the direction of the journey to work and he effectively demonstrated the existence of sectoral patterns in Liverpool and Manchester.

The most complete information on the journey to work for any British conurbation is that obtained for London in the 1951 Census. Not only was information on the usual residence and workplace of occupied persons collected for every household, but it was published for all the hundred-odd local-authority areas making up the conurbation.† The patterns are therefore analysable in some detail, certainly in much greater detail, paradoxically, than is possible for any other conurbation where a large fraction of the whole inner area always lies within one large local-authority area.

To test the predictions of the theory the pattern of the journey to work in London in 1951 is investigated in this chapter and the next. In this chapter we investigate, firstly, the pattern of the journey to work *to* two employment subcentres, and secondly the pattern of the journey to work *from* three residential areas. The areas were selected on the basis of the conclusions of Westergaard's investigation of the spatial distribution of employment in the London area. He found that, as well as the CBD (or Central Area, as it is designated in the Census), there are a few clearly marked secondary employment areas whose demand for labour considerably exceeds the local supply: there are the East End and southern riverside boroughs

† The 10 per cent sample Census of 1961 is known to be biased. In the 1966 Census this bias was reduced but, following the reorganisation of London government, information was published in respect of the new London boroughs which were usually two or three times as large as the old local-authoriy areas.

adjoining the Centre, the Acton-Brentford industrial area immediately west of the old county of London, and pockets further out, as at Kingston and Hayes (Westergaard, 1957, p. 43). Except for Kingston, in the south-west, these employment centres can be clearly seen to lie due east and west of the CBD (Fig. 12.4). The only other local authority area within the conurbation where the number employed exceeds the number of residents in employment are the two semirural areas of Elstree and Waltham Holy Cross on the northern edge of the conurbation.

The two local-authority areas chosen to investigate the

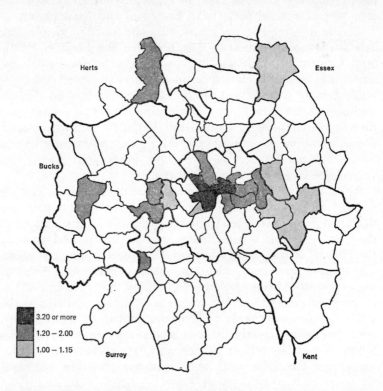

Fig. 12.4 Centres of employment in Greater London, 1951: job ratios by local-authority areas. The job ratio for a local-authority area is equal to the number of persons employed in the area divided by the number of employed persons living in the area. (See Fig. 9.3 for names of the local-authority areas.)

pattern of the journey to workplaces within their boundaries were Acton and Stepney. These were chosen because each is in one of the two main subcentres described by Westergaard, Stepney in the East End adjacent to the eastern boundary of the CBD, and Acton in the industrial area adjacent to the western boundary of the county of London. For both Acton and Stepney, the proportion of the occupied male population of every other local-authority area working in each of the two boroughs was calculated, and the results mapped. The patterns of residential location of the workers in each of the two boroughs are shown in Fig. 12.5 and 12.6.

To make the maps comparable, an index of commuting was constructed so that the value of the index in respect of

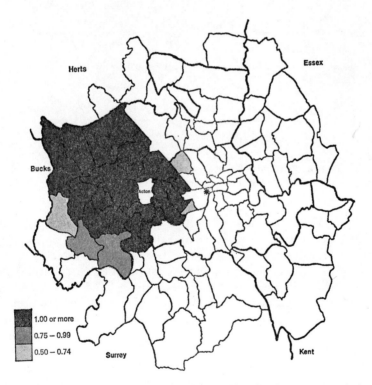

Fig. 12.5 Residential location of males working in Acton, 1951: index of commuting to Acton by local-authority areas. (See p. 206 for method of calculation of index, and Fig. 9.3 for names of the local-authority areas.)

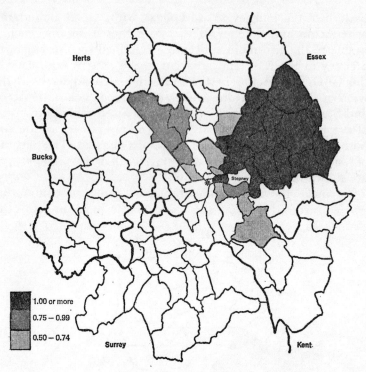

Fig. 12.6   Residential location of males working in Stepney, 1951: index of commuting to Stepney by local-authority areas. (See p. 206 for method of calculation of index, and Fig. 9.3 for names of the local-authority areas.)

travel from any local-authority area to a workplace in any other was independent of both the total occupied population living in the area of residence and the total working population in the area of work. This was done by calculating the proportion of the occupied resident male population in any local-authority area which worked in Acton or Stepney, and then dividing this by the proportion of the occupied male population of the Greater London conurbation which worked in Acton or Stepney. For example, 0·45 per cent of the occupied male population of Westminster works in Stepney and 3·82 per cent of the occupied male population of Dagenham. Since 2·20 per cent of the occupied male population of the conurbation work in Stepney, the index of commuting from Westminster to Stepney equals 0·45/2·20, or 0·2, and that for commuting from

Dagenham to Stepney equals 3·82/2·20, or 1·7. In Figs. 12.5 and 12.6 the areas shaded dark-grey have indices greater than or equal to 1·00, those shaded medium-grey have indices greater than or equal to 0·75 but less than 1·00, and those shaded light-grey have indices greater than or equal to 0·50 but less than 0·75. Unshaded areas within the conurbation have indices less than 0·50. Indices were not calculated for areas outside the conurbation boundary.

For Acton, the pattern of location is in accordance with the predictions of the theory. The map shows that those working in Acton tend to live in the same sector of the conurbation as their place of work. The results for Stepney are less clear, but the general pattern is as predicted: most of those working in Stepney live in the sector of London in which they work. Because Stepney lies closer to the CBD than Acton, far fewer workers travel outward to work in Stepney than travel outward to work in Acton. An exception to this rule arises because of the number of workers who travel from the north London boroughs of Hendon, Finchley, Hampstead, and St. Marylebone across the inner suburbs to work in Stepney. This appears particularly odd since this group of boroughs is separated from the north-eastern sector by boroughs which send very few workers to Stepney. This pattern would appear to be an effect of social agglomeration. However, whereas in Chapter 8 we attributed social agglomeration to the desire of people in the same income group to live close to each other, here the social agglomeration is due to the desire of people of the same race, culture, and religion to live close to each other. Typically Stepney was the first home of the Jewish immigrants to England, both because it was near the docks and because – for several reasons, but particularly because until 1832 Jews could not open retail shops in the City of London – it had been the traditional home of the Jews since the seventeenth century (see Hall, 1962, pp. 61f.). Even in 1930 it is estimated that 'nearly 30 per cent of the Jewish population of Greater London was still in Stepney and 56 per cent in inner East London, including Hackney and Stoke Newington. After 1945 Lipman doubts whether there were more than 30,000 Jews left in inner East London, out of some 280,000 in Greater London' (Hall, 1962, p. 62). Many, if not most, of the Jews who left east London would have moved on because of increased incomes, and thus gone to live in the higher-income Jewish areas. These are located in the boroughs of St. Marylebone, Hampstead, Finchley, and Hendon, the most well known being Golders Green in the south-east corner of the borough of Hendon. Even though this explains the pattern of residential location of those working in Stepney, it is still necessary to explain the location of the Jewish sector in the north rather than the east.†

The most plausible hypothesis would appear to be this. Initially Jewish immigrants settled in Stepney and both lived and worked there, particularly in the clothing industry. Later generations were more skilled than their parents, spoke English as their native language, and might have inherited some capital. They were therefore more mobile, and in many cases might move into industries located in the West End.

Thus, Lipman (1954) states that from the middle of the eighteenth century, at least, Jews had settled west of Temple Bar, to open retail shops, especially those serving the fashionable population of the then West End, goldsmiths, jewellers, watch-makers, embroiderers, etc. The location of the Court and Parliament in Westminster ensured that the rich and fashion-able lived in the West End. As a result, 'there were also many Jews living in St. Marylebone and Westminister who had moved from the City, or farther east, not because of business or trade reasons, but because they wished to reside in a fashion-able neighbourhood' (p. 14f.). Lipman (p. 76) quotes a con-temporary assessment of the position of the Jews in London to the effect that 11,000 of the 46,000 Jews in London in 1882 were in the families of merchants and professional people and, of these, 6,600 lived in the West End and west London and only 4,400 in north, south, or east London.

† One possible explanation was suggested by an American colleague. The location of the Jewish sector might be due to residents of the eastern suburbs being unwilling to sell homes to Jews. The only evidence I can discover in support of this explanation is the statement of a resident of Woodford quoted by Willmott and Young (1960): 'Half the people round here are from the East End. I came from Bow to East Ham. Then I moved out here. . . . The road where I was in Bow as a boy all did the same sort of thing. It was on account of the Jews. It was all Jews coming in and English moving out' (p. 4).

In the late nineteenth century and the early twentieth century London Jewry spread westwards, northwards, and eastwards from the two centres in the East End and the West End. The two main radial dispersions were due north from Stepney towards Tottenham and north-west from Westminster through St. Marylebone and Hendon. Of these two sectors, the north-western sector is clearly the more fashionable area, and as we have seen, many Jews with workplaces in Stepney are willing to travel across north London to live there. The map probably depicts a transitional stage, however. Over time we would expect later generations of Jews to have fewer business connections with Stepney, and hence, for the number of journeys between Hendon, etc. and Stepney to diminish. The

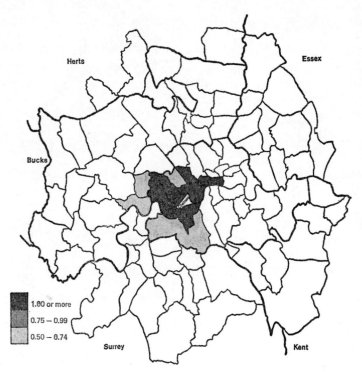

Fig. 12.7 Location of workplaces of those living in Chelsea, 1951: index of commuting from Chelsea by local-authority areas. (See p. 212 for a description of the method of calculation of the index, and Fig. 9.3 for names of the local-authority areas.)

1961 Census data shows just this. The number of males commuting to Stepney from Hendon, Finchley, Hampstead, and St. Marylebone fell by 19 per cent between 1951 and 1961, though the number of males working in Stepney fell by only 8 per cent.

As a further test of the predictions of the theory with respect to the pattern of the journey to work of those living in a particular area, three local-authority areas were selected, and three maps drawn showing the pattern of location of the *workplaces* of those living in the three areas. The local-authority areas were deliberately chosen so that there was no major employment subcentre between them and the CBD. If the theory is correct, workers should be willing to travel inwards

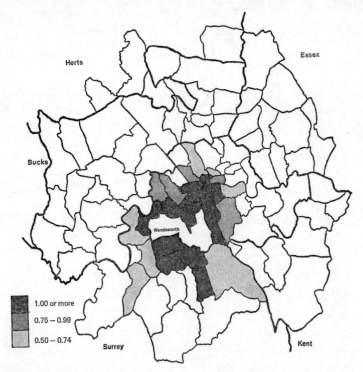

Fig. 12.8 Location of workplaces of those living in Wandsworth, 1951: index of commuting from Wandsworth by Local-authority areas. (See p. 212 for a description of the method of calculation of the index, and Fig. 9.3 for names of the local-authority areas.)

along radial routes as far as the CBD, but no further. The amount of outward commuting should be smaller, the closer the residential area is to the centre, for – as we showed at the end of the first section of this chapter – workplaces near the centre will draw very few workers from residential areas between them and the centre, while workplaces near the edge will draw a large proportion of their workers from suburbs between them and the CBD.

The patterns of the journeys to work of occupied males living in the local-authority areas of Chelsea, Wandsworth, and Merton and Morden are shown in Fig. 12.7, 12.8, and 12.9 respectively. The index of commuting is here mapped with respect to places of residence, whereas earlier they were mapped

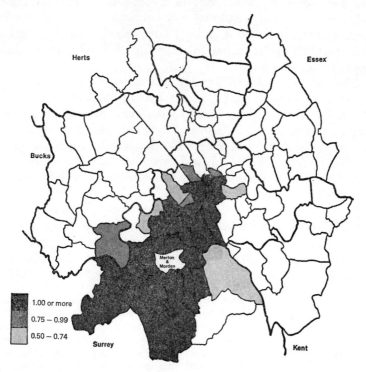

Fig. 12.9 Location of workplaces of those living in Merton and Morden, 1951: index of commuting from Merton and Morden by local-authority areas. (See p. 212 for a description of the method of calculation of the index, and Fig. 9.3 for names of the local-authority areas.)

with respect to places of work. Thus, since the index of commuting from Chelsea to Acton is equal to 0·56, Chelsea is shaded light-grey in Fig. 12.5, and Acton is shaded light-grey in Fig. 12.7. The index was calculated by finding the proportion of the male population working in any local-authority area and living in Chelsea, Wandsworth, or Merton and Morden, and then dividing this by the proportion of occupied male population of the Greater London conurbation living in Chelsea, Wandsworth, or Merton and Morden. It can easily be verified that the value of the index will be exactly the same whether it is calculated by this method or the method used earlier.† On the maps, the dark-grey areas are those for which the index of commuting to the named local-authority area is greater than or equal to 1·00; for medium-grey areas the index is less than one but greater than or equal to 0·75; for light-grey areas the index is less than 0·75 but greater than 0·50. Areas (lying within the conurbation) for which the index is less than 0·50 are left blank.

The patterns shown by the maps tend to confirm the theoretical predictions. Chelsea is a small borough adjacent to the CBD, with 0·5 per cent of the resident occupied male population of the conurbation. From Fig. 12.7 it is clear that in general those travelling to work towards the centre travel no further than the CBD. Outward travel is confined to short trips into the adjacent boroughs, with some longer outward journeys to the Acton-Brentford industrial area.

Wandsworth is a considerably larger borough than Chelsea, both in area and population (3·7 per cent), and lies somewhat further from the centre but in the same, south-easterly direction. Figure 12.8 shows that those commuting to the north of the borough tend to move along the radial routes towards the

† Let T_{ij} be the number of persons commuting from the ith local-authority area to the jth; let O_i be the occupied male population *living* in the ith area, and D_j be the occupied male population *working* in the jth area, and let P be equal to the total occupied male population living and working in the Greater London conurbation. Since the amount of commuting into and out of the conurbation is small, and can be ignored, $P = \Sigma O_i = \Sigma D_j$. If I is the value of the index of commuting from i to j:

$$I_{ij} = \frac{T_{ij}}{O_i} \bigg/ \frac{D_i}{P} = \frac{T_{ij}}{O_i} \bigg/ \frac{O_i}{P_i} = \frac{T_{ij} P}{O_i D_j}.$$

CBD. Very few have jobs in areas beyond the CBD. Journeys tend to be rather longer than those from Chelsea.

The third local-authority area, Merton and Morden, lies slightly further out than Wandsworth – once again in the south-east sector – and has 0·9 per cent of the occupied male population of the conurbation. Again, those travelling to work in the direction of the centre generally travel no further than the centre. In accord with the predictions of the theory, journeys outward are longer than is the case with the other two local-authority areas.

A STOCHASTIC MODEL OF THE JOURNEY TO WORK

One obstacle to our concluding that the patterns of journeys to work from local-authority areas shown in Figs. 12.7, 12.8, and 12.9 are in accord with the predictions of the theory is that the theory is deterministic, not stochastic. The predictions made in the first section of this chapter were developed using the assumption that there were only two (or at least a very few) defined workplaces in the city and that the boundaries of the 'commuting field' of any workplace or residential area could be clearly defined. While this model may hold if we are prepared to consider only the journey to work to the major subcentres of the city (like Stepney or Acton) it is not designed to provide a very good prediction of the pattern of the journey to work in a sector of the city like the south-east sector of London where in 1951 there were no very important employment subcentres, but where employment was fairly evenly distributed. In this case the theory should predict very little outward commuting whereas the empirical evidence shows that there is more than might be predicted.

Now it is evident that the theory we have used in this chapter (and in this book) is necessarily a simplification of reality. The location of the residence of a household is not solely determined by the location of the places of work of those members of the household who are working. Other factors must also determine the location of the household, e.g. ties of family or friends, the location of consumer services, the location of previous workplaces, the availability of jobs, the cost of moving, etc. All these factors can, however, be assumed to be random, since they are as likely to pull the household in one direction as in another. What we would wish to do therefore is to define a stochastic model of the journey to work which makes the maximum use of the results obtained from the (deterministic) theory, and which is maximally non-committal with respect to missing information. Wilson (1967) has shown that this most probable distribution of journeys is a variation of the 'gravity model':

$$T_{ij} = a_i . b_j . O_i . D_j . f(C_{ij})$$

where $f(C_{ij}) = e^{-\beta C}{}_{ij}$, and T_{ij} denotes the number of trips originating in zone i, with destinations in zone j (here the number of occupied males living in area i and working in area j), a_i and b_j, are constants applicable to the zones of origin and destination, respectively, which account for the numbers of trips between all zones besides zones i and j. O_i is the number of trip origins at zone i (here the number of occupied males in area i), D_j is the total number if trip destinations in zone j (here the number of males working in area j), and f is a decreasing function of C_{ij}, the cost of travel between zone i and zone j.

Usually travel time or distance will be used as a proxy for C_{ij}, but our theoretical investigations suggest that this is incorrect and the cost of travel must allow for the systematic variation of rents and wages within the city. For simplicity we shall ignore the existence of important employment subcentres and assume that density of employment is a function of distance from the centre of the city. In these circumstances the rent surface will be approximately conical in shape, with the steepness of the surface of the cone varying inversely with distance from the city centre. The wage gradient is also expected to have this shape. The financial cost of travel may be expected to vary uniformly with distance from the place of residence. If this model is correct, the cost of separation of residence and workplace varies with the distance of the place of residence from the CBD and the relative location of residence, workplace, and CBD.

If the workplace is located between the place of residence and the CBD the cost of separation is low because the increase in wages towards the centre reduces the direct cost of travel. On the other hand, if the CBD is located between the workplace and the place of residence the cost of separation may be high and will increase very rapidly as the distance of the workplace from the CBD increases, since the wage receivable will fall very rapidly.

The cost of residence–workplace separation if the residence lies between the workplace and the CBD – will depend on the distance of the place of residence from the CBD. If it lies close to the CBD, travel outward will be very costly because both the rent surface and the wage surface will be falling very steeply. If it lies near the periphery the two surfaces will be nearly flat, and the cost of residence–workplace separation will be only slightly greater than the cost of travel.

In the case of lateral travel the cost of workplace–residence separation will be approximately equal to the cost of travel, since the rent and wage surfaces will be more or less flat. By this form of analysis one can determine the most probable distribution of work trips for any area of residence. Two possible patterns are shown in Figs. 12.10(a) and 12.10(b), one for a residential area near the centre, the other for an area towards the periphery. It can be seen that they approximate to the actual patterns in Figs. 12.7 and 12.9 for Chelsea and Merton and Morden, respectively. The deviations appear to be due, firstly, to the radial pattern of transport routes which distorts the cost of radial travel relative to lateral travel so that in each case there would appear to be more outward travel and less circumferential travel relative to lateral travel than one would expect, and secondly, in the case of Chelsea, to the existence of the major employment subcentre in the Acton/Brentford area.

Major centres beside the CBD will tend to cause minor peaks in the rent and wage surfaces, thus causing the cost of residence–workplace separation to depend not only on the locations of the workplace and area of residence relative to the CBD, but also on their locations relative to the major sub-centres. Consideration of the earlier results derived from the theory will show the way in which the predictions of the stochastic model must be altered.

To conclude this section it may be noted that the analysis presented in this chapter provides an explanation for the

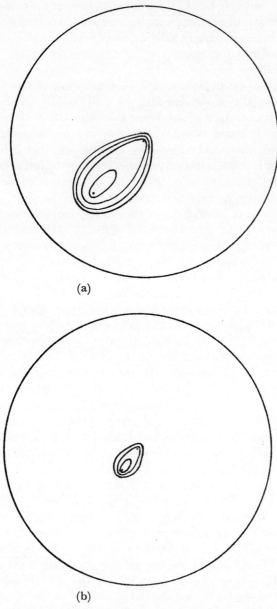

(a)

(b)

Fig. 12.10 Patterns of the journey to work from two representative residential areas.

discovery by Clark and Peters (1965) that the intervening opportunities model satisfactorily 'explains' the pattern of journeys to work between local-authority areas in London. They found that the relationship between the number of jobs within any given distance from a residential area and the percentage of the occupied residents of the area working within that distance appeared to be approximately the same for all local-authority areas studied.

Because of the increase in density of employment towards the centre, the number of opportunities (jobs) within a given distance from a residential area will be positively correlated with the distance of that area from the CBD. This can be put another way. The further a residential area is from the CBD, the greater must be the radius of any circle centred on the area which is drawn to circumscribe a given number of opportunities. Hence, the movement–opportunities relationship will appear to be more or less the same for all areas, even though the average distance travelled to work from each borough varies. It may thus be true to assert, with Clark and Peters, that 'distance does not matter'. But this is only true because the complex urban system adapts itself to ensure that workers travel from all parts of the city towards the CBD. Two 'unseen' mechanisms, the wage surface and the rent surface, and an empirical fact, the job-density surface, operate to modify the effects of the cost of travel alone and to ensure that the 'intervening opportunities' model works.

CONCLUSION

In this chapter we have tested the theory only with respect to employment centres other than the CBD. In Chapter 13 we take the analysis a stage further by dropping the assumption that the CBD is located at a single point, and analyse theoretically and empirically the pattern of journeys to work to a large CBD.

13 The Journey to Work – II: The Central Business District

In the preceding chapter we investigated the pattern of the journey to work in employment centres outside the central business district (CBD). In this chapter we carry out further theoretical and empirical analysis of the pattern of the journey to work. We shall show that, where the CBD is spatially extended, it is not true to say that each part of the CBD draws its workers from all areas of the city, even though it may be true that the CBD as a whole does draw its workers from all over the city. Thus, Carroll's conclusion that 'the residential distribution of persons employed in central districts tends to approximate that of the entire urban area population' (1952, p. 272) must be subject to qualification.

We shall also show that the desire to minimise location costs (the cost of the journey to work plus the rent) may conflict with the desire to reside in the same area as others of the same social class or income group and that the location of the household must therefore be determined through trading-off the costs and benefits of each possible course of action.

Central London is used as an example of a spatially extended CBD, and the theory is used to explain the complex pattern of the journey to work in the London CBD. In the first section some of the necessary theoretical analysis is developed.

THEORETICAL ANALYSIS

In the previous chapter it was shown that those working in employment centres outside the CBD will generally live in the same sector of the city as their workplace and on the opposite side of the workplace from the CBD. It requires only a minor extension of the theory to show that if the CBD is itself spatially extended then those working in each of its sectors will generally live in the same sector of the city as their workplace. We shall assume, as before, that all those working in the city have identical sets of bid-price curves, but shall suppose that the CBD consists of two spatially separate workplaces of equal size.

A cross-section through the equilibrium rent surface along the line of the two workplaces is shown in Fig. 13.1(a). One workplace is located at K_1 and the other at K_2. The equilibrium-rent surface at the cross-section is indicated by the line $ABCDE$. The position of the optimal bid-price cones of those working at K_1 is indicated by the line $ABCG$, and that of the others by the line $FCDE$. It can be seen that no member of either set of workers can move to a more preferred position. In Fig. 13.1(b) one set of workers lives and works in one half of the city, and the other set lives and works in the other half.

This deterministic theoretical exposition implies that the above conclusion will be true no matter how close together the two workplaces are, but a more realistic stochastic explanation would take their proximity into account, in the manner outlined in the section on the gravity model in the previous chapter. Obviously, the closer the two workplaces are together the less the cost of location in the 'wrong' sector and the less likely it is that the sectors of residence and work of each household will coincide. Hence, it is more likely that the sectoral pattern of journeys to work will be observable when the spatial extent of the CBD of a city is large than when it is small. Furthermore, it is also easily seen that the sectoral pattern of journeys to a workplace is likely to be less clear when the workplace is part of the CBD than when it is a subcentre of the city. In the former case 'non-optimal' locations are less costly.

So far, it has been assumed that all the workers at each workplace have identical bid-price curves. This assumption will now be dropped and variations in tastes will be introduced. Suppose that there are two workplaces in the city – the CBD and a subcentre – and that at each of the workplaces there are three sets of workers, each set having different tastes. In equilibrium the pattern of location of the households would be as shown in Fig. 13.2(b), while a cross-section through the rent surface along the line of the two workplaces would be as shown in Fig. 13.2(a). Workers in group 1 have the steepest

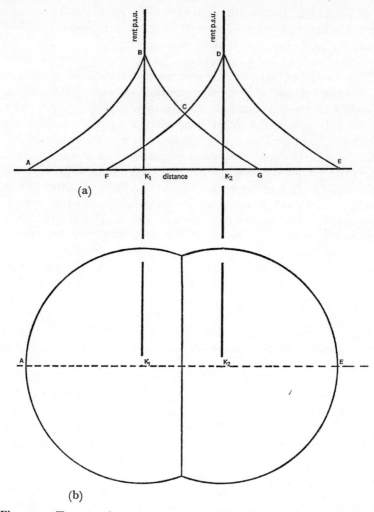

Fig. 13.1   Two equal centres: a cross-section through the rent gradient and the plan of the city.

bid-price cones, and locate close to their workplace, whichever it is. On the other hand, workers in group 3 have the least-steep bid-price cones, and always locate on the periphery, wherever their workplace lies. This will be true no matter how many workplaces there are in the city. Those households with the steepest bid-price cones will always locate near their

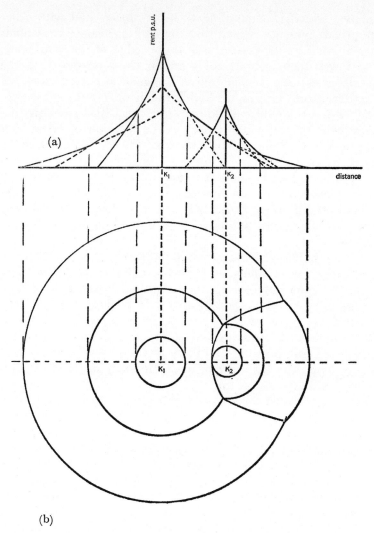

Fig. 13.2   The location of workers at a subcentre.

workplace, those households with the least-steep bid-price cones will locate on the periphery.

This result can be used to derive some empirically testable hypotheses. Thus, in Chapter 8 it was shown that in a city with a single workplace the pattern of location of households by income which is compatible with the theory and with other empirical evidence is such that the highest-income households are located at the centre and at the edge of the city, while the lowest-income households are in close proximity to the single workplace at the centre. In general, apart from the highest-income groups, the higher the income of the household, the further it would tend to live from the CBD. These predictions will be true with respect to the location of households about any workplace. The only qualification which must be made is that the number of high-income households which may wish to locate adjacent to an employment subcentre will almost certainly be too few to support a high-income neighbourhood, and these households will choose some other nearby location which can be shared with other high-income households.

Using the theoretical analysis in this chapter we can therefore state that low-income households will locate near the place of work, while high-income households may locate anywhere in the same sector as the workplace. In general, the lower the income of the household, the closer to the workplace it is likely to locate. Thus, we would expect that the low-income household living at the periphery of the city will be working there, while the high-income residents may be working anywhere in that sector of the city.

This theoretical analysis can be turned into an empirically testable proposition. For any workplace the proportion of the occupied residents of any residential area commuting to that workplace will tend to decline as the distance between the workplace and the area of residence increases. The rate of decline will depend on the incomes of the residents of the area. The rate of decline will be very high in the case of low-income residents, and very low in the case of high-income residents, i.e. the rate of decline will be inversely related to the incomes of the residents.

The same method of analysis can be used to explain the pattern of location of the female labour force. In Chapter 9 it was shown that, in a city with a single workplace, households in which a high proportion of the members worked would tend to locate close to the central workplace, while those with a low proportion working would tend to locate towards the

edge. From this we could deduce that the proportion of the
female population resident in any area who were in employment would tend to decline as the distance of the area from
the city centre increases. When there are many workplaces
(provided that we can make the simplifying assumption that
all members of a household work, if at all, at the same workplace)
households in which a high proportion of the members work
will live closer to the workplace than those in which a low
proportion work.† The females working at the CBD will tend
to live close to the CBD, and females working at subcentres
will tend to live close to these subcentre. In other words,
those females who are in employment and live at the periphery
of the city are likely to be employed there. This result can
be phrased as an empirically testable proposition. For any

† If this simplifying assumption is not made, the problem becomes much
more complex because the bid-price cone becomes a bid-price surface.
When there are two members of the household working at different workplaces, a cross-section through the representative surface would be as in
Fig. 13.3. The surface itself would look rather like two bell tents with
canvas covering the space between them. The optimal location of the
household will depend on the relative valuations of travel time of each
worker. The household may locate between the workplaces or outside
either of the two workplaces. Thus, in this case, an optimal location is
compatible either with outward commuting or with commuting across the
CBD.

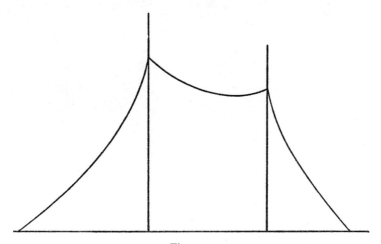

Fig. 13.3

workplace the proportion of the occupied males and females of any residential area will decline as the distance between the workplace and the area of residence increases, but the rate of decline will be greater for females than for males.

THE JOURNEY TO WORK IN THE LONDON CBD

In Fig. 12.5 the major employment subcentres of London were shown to be situated to the east and west of the CBD. The older industrial areas of the East End and the docks lie along the Thames to the east, while the newer industrial areas of Middlesex lie to the west. The maps of the residential locations

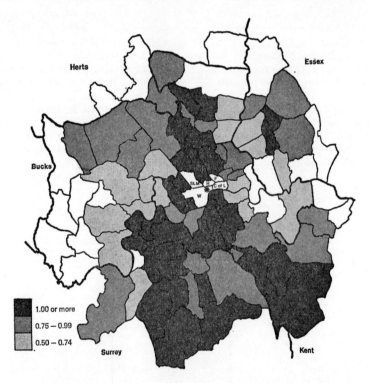

Fig. 13.4 Residential location of males working in five central local-authority areas of London, 1951: index of commuting to the five central local-authority areas by local-authority areas. (See p. 206 for method of calculation of index and Fig. 9.3 for names of the local-authority areas.)

of those working in Acton and Stepney (Figs. 12.6 and 12.7)
showed that the theoretical predictions made in Chapter 12
were confirmed. In general, those working in Acton lived in
the western sector of the conurbation, and those working in
Stepney lived in the eastern sector. If these patterns of location
can be assumed to be typical, it can be predicted that those
working in the CBD will tend to live in the northern and
southern sectors of the conurbation.

This is the 1951 pattern of location shown in Fig. 13.4 for
males and in Fig. 13.5 for females. Five local-authority areas,
the City of London, Holborn, Westminster, St. Marylebone,

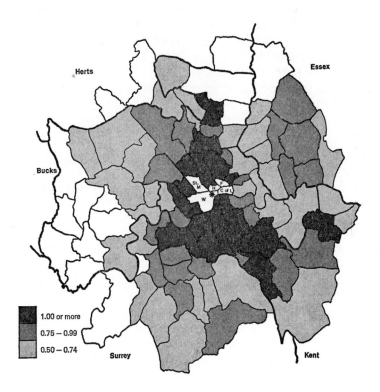

Fig. 13.5   Residential location of females working in five central local-
authority areas of London, 1951: index of commuting to the five central
local-authority areas by local-authority areas. (See p. 206 for method of
calculation of index and Fig. 9.3 for names of the local-authority areas.)

and Finsbury, were taken as representing the CBD.† The index of commuting between these five and the other local-authority areas was calculated for males and females separately, using the method explained in Chapter 12. An index greater than 1·00 indicates that more than 24·5 per cent of the occupied males resident in the area and more than 27·9 per cent of occupied females worked in the five central local-authority areas.

The influence of the major employment subcentres is clearly seen in the pattern of location of those working in central London. In the east and north-east relatively few travel to the centre but instead work in the nearer industrial areas of Essex and the East End. Similarly, in the west and south-west the residents tend to work in the Acton-Brentford area, Hayes and Harlington, or Kingston instead of travelling to the centre. This appears to be true for both males and females, even though the maps show that females working in the CBD tend to live in the inner areas of the conurbation, with a generally shorter journey to work than males.

It is interesting to compare the pattern of location of males working in the CBD shown in Fig. 13.4 with Fig. 13.6 which indicates the social class or economic status of the areas. The similarities between the locational pattern of central-area residents and the locational pattern of males in social classes 1 and 2 are remarkable but they are also predictable. The social structure of the work force in the CBD is very different from the average for the conurbation. Of the males working in the conurbation who are either employers, managers, or operatives in social classes 1 and 2, 42 per cent work in the five central local-authority areas, while only 20 per cent of the rest work in the centre. This can be put in another way: 35 per cent of those working in the five central local-authority areas are employers, managers, or operatives in social classes 1 and 2, while only 16 per cent of those working in the rest

† A small part of the built-up area of St. Marylebone lies outside the 'Commercial and Administrative Area' defined in the *Report on Greater London and Five Other Conurbations* (General Register Office, London, 1957) and a part of St. Pancras lies within that area. Since St. Pancras lies to the north of the centre, the latter omission probably leads to an understimate of the importance of the CBD as a workplace for residents in the northern sector.

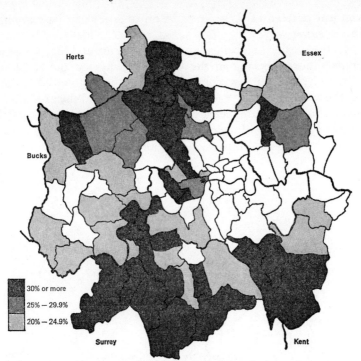

Fig. 13.6 Geographical distribution of social class, 1951: proportion of the occupied male population on social classes 1 and 2 by local-authority areas. *Source:* Census 1951, England and Wales, Occupation Tables (General Register Office London, 1957).

of the conurbation are in these categories. It follows that the patterns of location shown in Fig. 13.4 and 13.6 are likely to be similar. The local-authority areas where those working in the central area tend to locate are also likely to be areas with a high proportion of persons in social classes 1 and 2. In London, therefore, the differing social structures of employment centres may determine the differing social structures of residential areas; in particular, the pattern of location of employment centres will tend to cause a sectoral pattern of high status residential areas.† Investigation may show that the

† The pattern of location of households by income is approximately the same as the pattern of location of households by social class. This can easily be verified by comparing Fig. 13.6 with Fig. 5.7. of the London Traffic Survey (1964).

sectoral pattern of these areas in other cities can be explained in the same way.

In the earlier theoretical analysis it was argued that we would expect those working in the CBD to locate on the same side of the city as their workplace. Maps of the residential location of those working in each of the five central local-authority areas show that in general this is so. Figures 13.7 and 13.8 for Finsbury and St. Marylebone, respectively, show that most of those working in Finsbury† appear to live in the north and

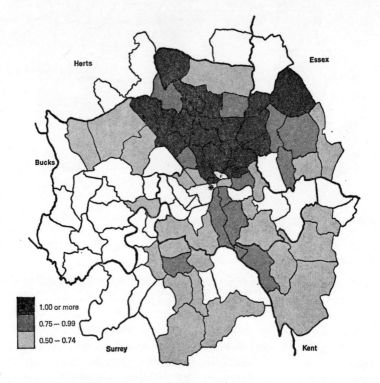

Fig. 13.7   Residential location of males working in Finsbury, 1951: index of commuting to Finsbury by local-authority areas. (See p. 206 for method of calculation of index and Fig. 9.3 for names of the local-authority areas.)

† The proportion of the occupied male population of the Greater London conurbation who work in each of the five central local-authority areas is as follows: Westminster 9·7 per cent, the City of London 8·1 per cent, St. Marylebone 2·7 per cent, Holborn 2·3 per cent, and Finsbury 1·8 per cent.

east, while most of those working in St. Marylebone live in the north and west. Holborn, the borough nearest the centre of the CBD, draws its workers mainly from the northern and southern sectors of the conurbation, as one would expect (map not reproduced here).

In the case of Westminster (Fig. 13.9) and the City of London (Fig. 13.10) there are only glimpses of the sectoral pattern. Thus, very few of those working in Westminster commute across the CBD from the east. Similarly, very few of those working in the City of London commute across the CBD from the west. From the maps it would appear that, while both areas draw most of their workers from the north and south,

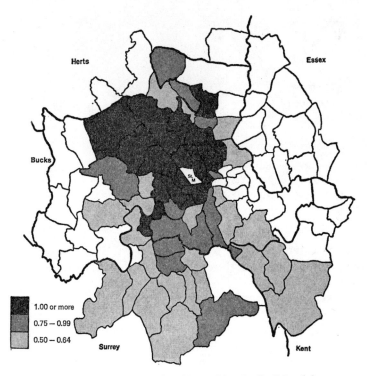

Fig. 13.8   Residential location of males working in St. Marylebone, 1951: index of commuting to St. Marylebone by local-authority areas (See p. 206. for method of calculation of index, and Fig. 9.3 for names of the local-authority areas.)

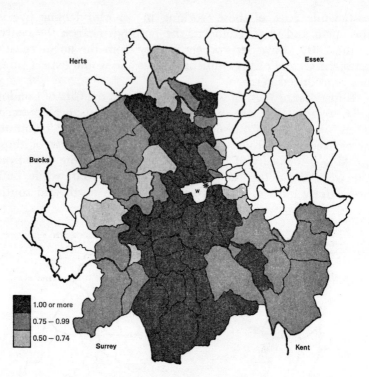

Fig. 13.9   Residential location of males working in Westminster, 1951 :
index of commuting to Westminster by local-authority areas. (See p. 206 for
method of calculation of index and Fig. 9.3 for names of the local-authority
areas.)

there is a tendency for those working in Westminster to live
towards the west, and those working in the City to live towards
the east.

What is quite clear from the maps is that, while the theoretical prediction that persons will live in the same sector of the
conurbation as their work is only partly confirmed, the evidence
definitely refutes the usually accepted belief that 'the residential
distribution of persons employed in central districts tends to
approximate that of the entire urban population' (Carroll
1952, p. 272).

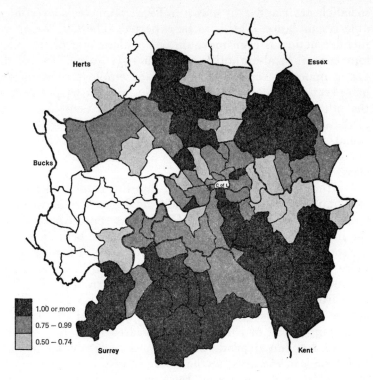

Fig. 13.10   Residential location of males working in the City of London, 1951: index of commuting to the City of London by local-authority areas. (See p. 206 for method of calculation of index, and Fig. 9.3 for names of the local-authority areas.)

THE PATTERN OF LOCATION OF THE CITY OF LONDON'S WORKERS

From the maps it is evident that the residential distribution of those employed in the City of London is the nearest approximation to that of 'the entire urban population'. In this section we shall use econometric methods to determine whether or not there is a tendency for those working in the City to live in the east.

Comparison of the pattern of location of the City's male workers (Fig. 13.10) with the pattern of location of males in social classes 1 and 2, as shown in Fig. 13.6, reveals a considerable resemblance. This is to be expected. The City's employment structure is even more untypical than that of the other four central local-authority areas. In 1951 40 per cent of the City's male workers were classified as employees, managers, or operatives in social classes 1 and 2, against 33 per cent in the other four central boroughs and 16 per cent in the rest of the conurbation. Thus, a possible explanation for the locational pattern of the City's male workers is that they choose to locate in the same areas as others of the same social class.

On the other hand, it is evident that this cannot be the only explanation, for it does not explain the differences between the two patterns of location. The City draws more workers from the inner areas of the conurbation than would be indicated by the social class of those living in the areas, and it tends to draw more workers from the north-east than would be indicated by the social class of those living in north-eastern boroughs. Both the cartographic evidence and the theoretical analysis set out earlier therefore suggests that the proportion of the occupied male population of a local-authority area commuting to the City of London depends firstly, on the proportion in each social class in the area, secondly, on its distance from the City of London, and thirdly, on the direction of travel from the City to the area.

Setting aside for the moment the dependence of the proportion commuting on direction, a plausible form of regression equation reflecting the structural relationship would be:

$$C = (a_1 + b_1 . \log D) U + (a_2 + b_2 . \log D) M + (a_3 + b_3 . \log D) L$$

$$(13.1)$$

where C is the percentage of the occupied male population of a local-authority area who work in the City of London;

U is the percentage of the occupied male population of a local-authority area in social classes 1 and 2;

M and L are the percentages of this same segment of the population in social classes 3, and 4 and 5, respectively;

$U + M + L = 100$; and

D is the (airline) distance from the City of London to the local-authority area.†

In this form the equation states that the proportion of the male occupied population of any local-authority area who work in the City is equal to the sum of fractions of the proportions of the occupied male population in each social class, and that these fractions will vary with distance from the local-authority area to the City. We would expect the fractions to decline with distance, and from the earlier theoretical analysis we would expect the rate of decline to be least for the highest-income groups and greatest for the lowest income groups, i.e. we would expect $0 > b_1 > b_2 > b_3$.

Equation (13.1) cannot be used to fit a regression because $U + M + L = 100$. If we substitute $100 - U - M$ for L in (13.1) the equation becomes:

$$C = A_1 + B_1 . \log D + A_2 . U + B_2 . U . \log D + A_3 . M + B_3 . M . \log D \tag{13.2}$$

where $A_1 = 100a_3$, $B_1 = 100b_3$;
$A_2 = a_1 - a_3$, $B_2 = b_1 - b_3$;
$A_3 = a_2 - a_3$, $B_3 = b_2 - b_3$. We would expect to find that $B_2 > B_3 > 0 > B_1$.

A regression equation in the reduced form of (13.2) was fitted to the data relating to 114 local-authority areas, 94 within the conurbation and 20 outside it and adjacent. The direction of travel was taken into account through a system of dummy variables. The London region was divided into six sectors, each radiating from the City of London, and every local-authority area (except the City) was allocated to a sector. These sectors are shown in Fig. 13.11. In the regression equation SE, S, N, NW, and SW are dummy variables denoting the south-eastern, southern, northern, north-western, and south-western sectors respectively, and such that each dummy variable takes the value one if the local-authority area lies in the sector denoted, and zero if it does not. The coefficient of the

† It would have been preferable, on theoretical grounds, to use time or route distance instead of airline distance as an idependent variable. But it is known that both time and route distance are highly correlated with airline distance so that not much is lost by using the latter, while it is considerably easier to calculate. See, for example, Wilson (1967).

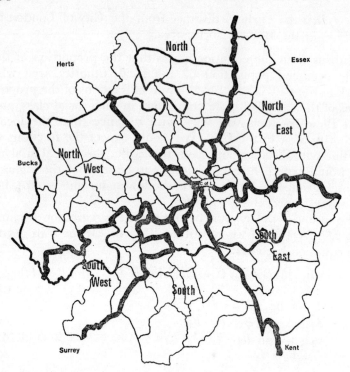

Fig. 13.11 Sectors of the Greater London region (radiating from the City of London).

variable which would otherwise denote the north-eastern sector is constrained to equal zero. The equation obtained was:

$$C = 3\cdot13 - 15\cdot19.\log D + 0\cdot10.U + 0\cdot33.U.\log D +$$
$$(5\cdot76) \qquad (0\cdot06) \quad (0\cdot06)$$

$$0\cdot13.M + 0\cdot08.M.\log D$$
$$(0\cdot09) \quad (0\cdot10)$$

$$-0\cdot79.SE - 2\cdot01.S - 3\cdot45.N - 4\cdot79.NW - 4\cdot86.SW \quad R^2 = 0\cdot83$$
$$(0\cdot44) \quad (0\cdot51) \quad (0\cdot45) \quad (0\cdot43) \quad (0\cdot52)$$

$$(13.3)$$

where the standard errors of each coefficient is shown in parentheses beneath the coefficient.

The coefficients of the dummy variables are negative in all

cases, and with the exception of the coefficient of SE, the coefficients of all the dummy variables are significant at the 1 per cent level. Thus, once differences in the social class structure of local-authority areas in the different sectors are allowed for, these negative coefficients indicate that people working in the City try to live in, or near, the north-eastern sector.

The coefficients of the distance variables take the expected signs. The negative coefficient of $\log D$ indicates that the proportion of social classes 4 and 5 in any local-authority area commuting to the City declines rapidly with distance. The positive coefficients of $U.\log D$ and $M.\log D$ and their order of magnitude indicate that the rate of decline in the proportion of those in social class 3 commuting to the City is less rapid than the rate of decline in the proportion of those in social classes 3 and 5 but is more rapid than the rate of decline in the proportion of those in social classes 1 and 2 commuting to the City. This result is in accordance with the theoretical predictions.

On the other hand, when the equation is put into its structural form it is seen that the proportion in social classes 1 and 2 commuting to the City actually appears to increase with distance, since the coefficient of $U.\log D$ is positive. For example, the structural form of the equation for the north eastern sector is

$$C = (0{\cdot}13 + 0{\cdot}18.\log D)U + (0{\cdot}16 - 0{\cdot}08.\log D)M + (0{\cdot}03 - 0{\cdot}15.\log D)L \qquad (13.4)$$

The two most probable explanations would seem to be these.

1. The proportion of the resident population in social class 3 varies very little between local-authority areas. In all except one area the proportion of the male occupied population in social class 3 lies between 40 and 65 per cent. As a result, $M.\log D$ and M and $\log D$ are highly correlated and this multicollinearity makes the estimates of the coefficients of these variables in (13.3) highly unreliable; in turn, this causes the coefficients of $U.\log D$, $M.\log D$, $L.\log D$, and M in (13.4) to be unreliable.

2. According to (13.4), the rate of decline in the proportion of those in social classes 4 and 5 commuting to the City is

so rapid that it declines to zero after 2 miles and is negative thereafter. Obviously, this is impossible but no constraint has been placed on the form of the regression equation to prevent it. It would appear that, in the regression, the increase in the negative proportion of L with distance is compensated by the increase in the positive proportion of U.

These two explanations are not necessarily alternative, and may be complementary. Obviously, some improvement in the form of the regression equation would possibly remove the second problem at least, but this would seem to be unnecessary in the present context when the aim is not extreme accuracy but a reasonably good fit which confirms the theoretical analysis, and this has been achieved.†

Firstly, the estimated coefficient of $U . \log D$ in (13.3) is positive and significantly different from zero at the 1 per cent level. This evidence strongly supports the theoretical prediction that the rate of decline in the proportion of the higher-income groups commuting to the City is lower than the rate of decline in the proportion of the lower-income groups.

Secondly, the estimated coefficients of the dummy variables in (13.3), and their level of significance, show almost conclusively that, after allowing for the particular social class structure of a local-authority area, and its distance from the City, the proportion of the occupied male population of the area commuting to the City depends on the direction of travel from the City to the local-authority area. An area to the east of London having the same social structure and being located the same distance from the City as a local-authority area in any other direction will have a higher proportion of its male occupied population commuting to the City.

† One improvement which would certainly raise the level of examination would be to increase the number of sectors. For example, if a regression equation in the reduced form of (13.2) is fitted to the data for the northern and north-eastern sectors, along with a single dummy variable denoting the northern sector, the coefficient of determination (R^2) equals 0.67, and the coefficient of the dummy variable is -1.83. If the northern and north-eastern sectors are split into four subsectors, and three dummy variables are used, the value of the coefficient is increased to 0.93 while the coefficients of the three dummy variables are of the expected sign (reading anti-clockwise, the coefficients are -2.61, -3.73, -5.25).

One problem does arise with respect to this latter conclusion. Workplaces in the employment sub-centre to the east of the City (i.e. the East End and docks) employ mainly low-income manual labour, while workplaces in the City employ mainly high-income non-manual labour. While it is plausible to argue that the proportion of the high-income residents in the eastern sectors in the City will be higher than the proportion of those in other sectors, it does not seem as plausible to argue that the proportion of the low-income residents in the eastern sectors working in the City will be higher than the proportion of those in other sectors, since we would expect them to work in the East End instead of the City.

It may be possible that the sectoral pattern indicated by the signs on the dummy variables in (13.3) is only true for those in social classes 1 and 2. To test this hypothesis five further interaction variables were added to the regression to discover whether there was any significant interaction between social class and the direction of travel. In part, the results obtained were:

$$C = \ldots + 0{\cdot}08{\,.\,}U{.}\text{SE} + 0{\cdot}04{\,.\,}U{.}\text{S} - 0{\cdot}05{\,.\,}U{.}\text{N} + 0{\cdot}02{\,.\,}U{.}\text{NW}$$
$$\quad (0{\cdot}07) \qquad (0{\cdot}07) \qquad (0{\cdot}07) \qquad (0{\cdot}05)$$
$$- 0{\cdot}04{\,.\,}U{.}\text{SW} \qquad\qquad (13.5)$$
$$(0{\cdot}08)$$

If the hypothesis were correct, the signs of the interaction variables would all be negative since the fall in the proportion of those in social classes 1 and 2 in any area commuting to the City from any sector, when compared with a similar area in the north-eastern sector, would be greater than for other social classes. Indeed, the proportion in the other social classes commuting to the City may actually increase away from the north-eastern sector. In fact, the signs of the coefficients of the interaction variables in (13.5) provide no evidence to confirm the hypothesis. Three of the signs are positive and two are negative. Furthermore, none of the coefficients is significantly different from zero. These results indicate that there seems to be no interaction between social class and direction. The proportion of those in social classes 3, 4, and 5 commuting to the City varies between sectors in the same way as the proportion of those commuting in social classes 1 and 2.

The most plausible explanation for the pattern of location
of the City's workers takes into account the desire of both high
and low-income workers to live among others with the same
incomes.

The City must be viewed as a major employment centre
located between two minor centres. To the west and north-
west of the City is the rest of the CBD, and this is a larger
employment centre than the City. To the east and south-east
of the City are the docks and the industrial area of the East
End, which are considerably less important as an employment
centre than the City. In common with the rest of the CBD,
the City has a mainly clerical labour force with a high median
social class. The East End, on the other hand, has a mainly
manual work force with a very low median social class. As a
result, as we noted earlier, local-authority areas with residents
of a high median social class tend to be located to the north
and south of the CBD, while local-authority areas to the east
generally have residents of a low median social class.

Consider any group of workers in social classes 1 and 2
living in any local-authority area a given distance from the
CBD. If the area lies to the north or south of the CBD most of
them will work in the CBD, mainly in the City or Westminster.
If, on the other hand, the area lies to the east, most of the
group will still work in the CBD, since there are few employment opportunities for them in the East End, but they will
tend to work in the City and not in Westminster. Thus, the
proportion of those in social classes 1 and 2 living in the eastern
sector who work in the City will be higher than the proportion
of those living in the northern, southern, or western sectors who
work in the City.

Now take those in social classes 4 and 5 who work in the
CBD. While location to minimise location costs alone would
presumably suggest location to the north and south of the CBD,
location in the eastern sector allows each worker to live in the
same area as large numbers of others in the same social class.
Hence, for those working in the City, location in the east is
advantageous. On the other hand, for those working in
Westminster, location in the east may mean paying too great
a cost in the form of a lengthened journey to work, so that a
greater proportion of those working in Westminster will locate

in the north, south, and west. Hence, of any group of residents in social classes 4 and 5 working in the CBD, the proportion working in the City will be higher if they live to the east than if they live in any other direction. Now, if the concentration of employment in the CBD is large relative to the East End employment centre, and the concentration of employment in the rest of the CBD is large relative to employment in the City, a larger proportion of those in social classes 4 and 5 and living in the eastern sectors will work in the City than those in the same social classes but living in any other sector.

Thus, the negative signs on the dummy variables in (13.3) occur not only because of the forces predicted in the theoretical analysis, but for two reasons which are to some extent fortuitous; firstly, because the employment structure of the CBD differs considerably from that of the East End, and secondly, because of the unimportance of the East End as an employment centre relative to the CBD.†

That the latter may be the only necessary condition is shown by the pattern of location of females working in the City. This is illustrated in Fig. 13.12. The directional pattern suggested by the coefficients of the dummy variables in (13.4) is clearly shown on the map. The highest proportion commutes from the north-east, the second highest from the south-east, the third highest from the south, the fourth highest from the north, and the lowest proportion from the western sectors. In the pattern of location of the female population it is apparent that differences in the social structure of the suburbs are largely irrelevant, and so the sectoral pattern shows up on the map.

The pattern of location of females working in the City serves also to refute any suggestion that the pattern of location of males working in the City might be determined by distortions of the radial transport system. The differences between the two patterns of location and the similarity of the pattern of location of females to the pattern of location of males, *if there were no differences in the social class structure of residential areas*, confirms that the locational pattern of the males is determined by the trade-off between a desire to minimise location costs by choosing

† A possible third reason may be the proximity of the East End subcentre to the CBD. This means that there is no great difference in the cost of working in one rather than the other.

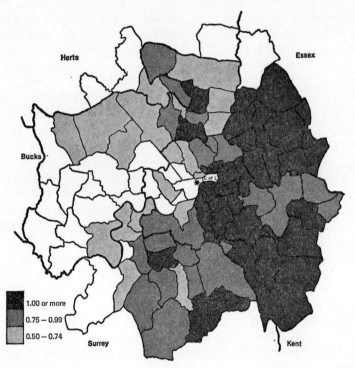

Fig. 13.12   Residential location of females working in the City of London, 1951: index of commuting to the City of London by local-authority areas. (See p. 206 for method of calculation of index, and Fig. 9.3 for names of the local-authority areas.)

a location in the workplace sector and a desire to locate in the same area as others of the same social class.

CONCLUSION

In this chapter we have replaced Carroll's 1952 conclusion by a more complex statement. The residential distribution of persons employed in the central districts need *not* tend to approximate that of the entire urban population. It will be determined by the location of the more important employment subcentres and by the differences in the income distribution among (or the social class structure of) those working in the central districts and those working in the employment subcentres.

14 Integrating the Theory: Variations in Household Size

In the later chapters of this book we have set out and tested each theoretical prediction separately. We have not considered them together. Yet it is obvious that if we want to use the theory to predict changes in the spatial structure of the city we must take into account the interactions between the various characteristics of households which we have so far considered separately.

As an example of these interactions we shall use the results set out in previous chapters to explain the variation in household size in the city more fully than we were able in Chapter 9. In doing so, we shall integrate the theoretical analysis of the pattern of location of households of differing size in Chapter 9, the analysis of the pattern of location of households with differing incomes in Chapter 8, and the analysis of the pattern of the journey to work in a city with many workplaces in Chapters 12 and 13.

THEORETICAL ANALYSIS

We showed in Chapter 9 that the householder in any particular income group will locate so that the smallest households tend to live nearest the centre, and the largest tend to live nearest the periphery. In Chapter 8 we showed that the households in any particular income group will tend to locate so that those with the lowest income elasticities of demand for space live nearest the centre, and those with the highest income elasticities live nearest the periphery. The resultant geographical distribution of income groups in the city is one in which the high-income groups are concentrated near the city centre and

on the city's periphery but are also scattered over the intermediate areas, while other income groups are concentrated
at distances from the centre which increase with income.

These results suggest that in a city with a single workplace
the average size of households in the highest-income groups
would tend to increase with distance from the city centre,
reaching a maximum at the edge, as indicated by line *AA* in
Fig. 14.1. On the other hand, the average size of households
in the lowest-income groups will increase with distance from
the centre but reach its maximum after only a short distance,
as shown by line *AB*, since households in the lowest-income
groups must locate close to the centre; no matter how large
the household, its low income will not allow it to occupy very
much space, and so it will never find it worth while to live
very far from the centre.

Suppose now that there are several workplaces in the city.
We showed in Chapters 12 and 13 that low-income households
will locate close to the subcentre in which the head of the
household works but that high-income households will tend to
locate over the area between the subcentre and the edge.
Suppose that, in addition to the city centre, there are two

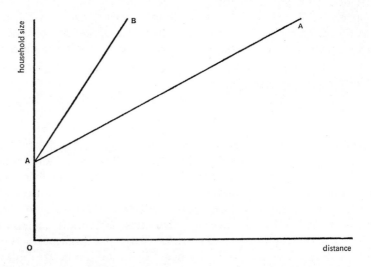

Fig. 14.1   The variation of household size with distance from the city
centre for two income groups.

subcentres. In Fig. 14.2, lines OA, $A'O'A$, and $A''O''A$ indicate
the expected relationship between average household size and
distance from the city centre for high-income households
whose heads of household work in the centre and subcentres.
The expected relationship between average household size
and distance from the city centre for low-income households is similarly indicated by the three lines OB, $B'O'B'$, and
$B''O''B''$.

It can be easily seen that if there is a multiplicity of workplaces in the city, then, taking all the households together
whatever the workplace of the head of household, the average
size of households in the low-income groups will not vary
noticeably with distance from the city centre. The probable
relationship is shown by line bb in Fig. 14.3. On the other
hand, the average size of households in the high-income groups
will tend to increase continuously with distance from the
centre, as indicated by the line oa. We would expect, therefore,
that the average household size in high-income neighbourhoods
in the inner areas of the city will be low, and in the outer
areas of the city it will tend to be high. Average household

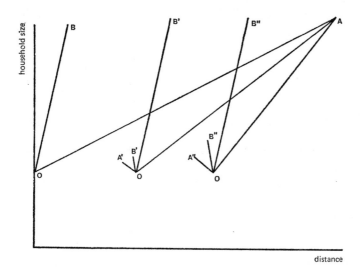

Fig. 14.2   The variation of household size with distance from the city
centre for two income groups when there are two subcentres.

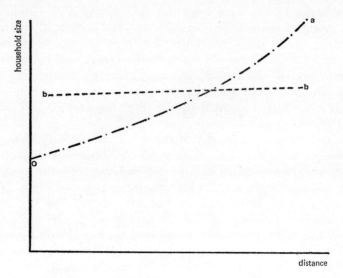

Fig. 14.3   Average household size as a function of distance from the centre of a city with many workplaces.

size in low-income neighbourhoods will vary very little with distance from the city centre.

VARIATION IN HOUSEHOLD SIZE IN LONDON AND BIRMINGHAM

From the above analysis it can be seen that the relationship between distance and household size is not a simple one. In Chapter 9 we regressed average household size against distance from the city centre, using data for seven large British cities. The results showed that there was generally a weak, but positive correlation between household size and distance, and these results were similar to those obtained by Muth, using data for sixteen U.S. cities. While these results were sufficient to confirm the theoretical prediction that there would be a positive correlation, they explained very little of the variation in household size between different areas of the city. The above analysis suggests that household size is a function of distance and of the income of the neighbourhood, with some interaction between the two independent variables. When British data are

used, information on incomes is not available but we can use
the proportion of the population of the area in the highest
socioeconomic groups as a surrogate. A possible regression
equation reflecting the structural relationship outlined above
would be:

$$H = (a_1 + b_1 . D)S + (a_2' + b_2' . D)R' \qquad (14.1)$$

where H is the average size of households in the ward or
borough;

S is the proportion of the occupied male population of
the ward or borough in socioeconomic groups
1, 2, 3, 4, and 13, as a fraction of 1;

R' is the proportion of the occupied male population of
the ward or borough *not* in socioeconomic groups
1, 2, 3, 4, and 13, as a fraction of 1;

$S + R' = 1$; and

D is the (airline) distance from the centre of the city to
the ward or borough.†

Since we intend to fit the equation to data for British cities
we should also take into account the fact that much of the
housing in British cities is owned by local authorities so that
the occupation of local-authority housing at any given location
is determined outside the market. Therefore, the average
household size in an area will depend on the proportion of the
population in local-authority housing, and there may be some
interaction between this proportion and distance, since there
is no reason to suppose *a priori* that local authorities will not
vary the size of the household they allocate to a dwelling with
the dwelling's distance from the centre. The structural form
of the regression equation then becomes:

$$H = (a_1 + b_1 . D)S + (a_2 + b_2 . D)L + (a_3 + b_3 . D)R \qquad (14.2)$$

where H is the average size of households in the ward or
borough;

S is the proportion of the occupied male population of
the ward or borough in socioeconomic groups 1, 2,
3, 4, and 13, as a fraction of 1; we assume that this

† It is known that both journey time and journey distance are highly
correlated with airline distance so that not much is lost by using the latter
and it is considerably easier to calculate.

is equal to the proportion of the households in the ward or borough in high-income groups;

L is the proportion of the households in the ward or borough living in housing rented from a local authority, as a fraction of 1; we assume that no households in high-income groups live in local-authority housing.

R is the remainder, i.e. the proportion of the households in the ward or borough not living in local-authority housing and not in high-income groups;

$S + L + R = 1$; and

D is the (airline) distance from the centre of the city to the ward or borough.

In this form, equation (14.2) states that the average size of the households in an area at a certain distance from the centre will depend on the average size of households in each of three categories (high-income, local-authority, and the remainder) at that distance from the centre, and on the proportion of the households in the area in each of the three categories. We would expect b_2 to be significantly greater than zero since we expect the average size of high-income households to increase with distance from the centre; we do not expect b_3 to be significantly different from zero since we do not expect the average size of low-income households to vary with distance from the city centre. Furthermore, since local authorities are likely to try to ensure a mix of household sizes in any housing estate that they own, we expect that the average size of households in local-authority housing will not vary with distance from the city centre, and so b_2 is not likely to be significantly different from zero.

Equation (14.2) cannot be used to fit a regression because $S + L + R = 1$. Substituting $1 - S - L$ for R in (14.2), we obtain:

$$H = A_1 + B_1.D + A_2.S + B_2.S.D + A_3L + B_3.L.D \qquad (14.3)$$

where $A_1 = a_3$, $B_1 = b_3$;

$A_2 = a_1 - a_3$, $B_2 = b_1 - b_3$;

$A_3 = a_2 - a_3$. $B_3 = b_2 - b_3$.

In this reduced form the equation can be used to fit a regression. From the above argument we expect that B_1 and B_3 will not be significantly different from zero.

In this form the equation was fitted to data for 1961 for the county of London and the county borough of Birmingham.†
The equations obtained were:

London

$$H = 2·70 - 0·00.D - 3·08.S + 0·42.S.D + 0·10.L + 0·19.L.D \quad R^2 = 0·92$$
$$\quad\quad (0·09) \quad (0·86) \quad (0·27) \quad\quad (0·74) \quad (0·42)$$

$$(14.4)$$

Birmingham

$$H = 2·99 + 0·07.D - 2·63.S + 0·24.S.D + 0·88.L - 0·12.L.D \quad R^2 = 0·80$$
$$\quad\quad (0·06) \quad (1·11) \quad (0·35) \quad\quad (0·22) \quad (0·08)$$

$$(14.5)$$

Note that a simple regression of average household size against distance explained only 0·33 of the variation in the county of London and virtually none of the variation in Birmingham, so that there is a considerable improvement in the proportion of the variation explained when the equations are more carefully specified.

The similarity between the coefficients in (14.4) and (14.5) is striking. As expected, neither the coefficient of D nor that of LD, is significantly different from zero in either equation. The coefficients of S are positive and highly significant in both cases. The coefficients of SD are positive, as predicted, and though neither is significant, they are both of the same order or magnitude. The coefficient of L is positive, indicating, as one might expect, that both the local authorities tend to house larger households in preference to smaller households. In the structural form of the equations, these resemblances are also brought out. The equations are:

London

$$H = (-0·38 + 0·42.D)S + (2·80 + 0·19.D)L + (2·70 - 0·00.D)R$$
$$(14.6)$$

† The observations were for local-authority areas in London and for wards in Birmingham. Except for distances, the data were those collected in the 1961 Census. The data for London are published. The data for Birmingham was made available to me by A. B. Neale, the Birmingham Corporation statistician, and is in respect of the wards which came into existence when the Birmingham ward boundaries were redrawn in 1962.

Birmingham

$$H = (0.36 + 0.31.D)S + (3.87 - 0.05.D)L + (2.99 + 0.07.D)R$$
$$(14.7)$$

It is clear that these results are in accord with the theoretical analysis. While the average size of high-income households is low in the city centre and increases rapidly towards the edge, the average size of other households varies very little with distance from the centre. The equations were therefore re-estimated in the form:

$$H = A_1 + A_2.S + B_2.S.D + A_3.L \qquad (14.8)$$

omitting D and $L.D$ from the equation. Theoretically, there is no reason why the coefficient of D and $L.D$ should differ from zero, and in these two cases we have shown that in fact they differ little from zero. Furthermore, there is considerable correlation between D, $S.D$, and $L.D$ so that inclusion of all three as independent variables results in considerable multi-collinearity, with a consequent reduction in the accuracy of the estimates of the coefficients.

When equations in the form of (14.8) are fitted to the data we obtain, in reduced form:

London

$$H = 2.72 - 3.43.S + 0.51.S.D + 0.63.L \qquad R^2 = 0.90 \qquad (14.9)$$
$$\quad\;\; (0.32) \quad\; (0.08) \qquad (0.22)$$

Birmingham

$$H = 3.27 - 3.78.S + 0.37.S.D + 0.54.L \qquad R^2 = 0.78 \quad (14.10)$$
$$\quad\;\; (0.72) \quad\; (0.15) \qquad (0.14)$$

It can be seen that there is virtually no reduction in the variance explained by either equation, so that the equation is rather more useful in this form than in the other. All the coefficients are significant and are again remarkably similar. Both the equations are fitted to data for the inner areas of cities, and not to data for the whole conurbation. The inner areas exclude the high-income suburbs of the cities, and those high-income households in the inner areas will tend to be small in size. As a result, the negative coefficients of S in (14.9) and (14.10) are highly significant, and variations in S explain most of the variation in household size. This might be expected from Fig. 14.3, since only the inner ends of *oa* and *bb* are being estimated. As a further test of the theory, equation 14.8 was fitted to the data for Greater London,† to obtain, in reduced form:

Greater London

$$H = 2 \cdot 90 - 3 \cdot 05 \,.\, S + 0 \cdot 28 \,.\, S \,.\, D + 0 \cdot 48 \,.\, L \qquad R^2 = 0 \cdot 77 \qquad (14.11)$$
$$(0 \cdot 28) \quad (0 \cdot 02) \qquad (0 \cdot 13)$$

In this equation, the relative importance of S and $S.D$ in explaining variations in household size is the reverse of that for the county of London. While both are highly significant, the interaction between income and distance is now more important than income alone.

One point should be made about the place of equation (14.8) in an econometric model of the spatial structure of the city. The spatial patterns are determined simultaneously and any econometric model must take account of this simultaneity. Now in (14.8), while L, the proportion of the population in local-authority housing, and D, the distance between the local-authority area and the CBD, can be assumed to be determined exogenously, S, the proportion of the population in the high-income groups is not an exogenous variable, but is determined within the system.

It might be argued, therefore, that the results obtained in equations (14.9), (14.10), and (14.11) will be biased. However, this is not so. In the model, S is determined before H is determined. While, to determine the average size of households in the area, we need to know the income distribution of those living in an area, we do not need to know the average household size to determine the income distribution. The system is therefore recursive and the estimates of the regression equations for average household size are unbiased.‡

† The observations relate to the ninety-eight local-authority areas existing in 1961, which lay within the conurbation defined in the Census or within the area which came under the control of the Greater London Council in 1965.

‡ Specifying an equation which would explain variations in S is more difficult than specifying one to explain variations in H, the main difficulty being that we have to allow for social agglomeration.

15 Policy and Planning

In this book we have developed a positive theory of residential location; we have been concerned with what the pattern of location is, not with what it ought to be. In Chapter 7, and elsewhere, we have discussed some of the normative aspects of variations in residential density, as has Mirrlees (1972), but a normative theory of residential location still awaits development. This does not mean, however, that we cannot use the positive theory to assess the feasibility or effects of policies or plans which will alter the spatial structure of a city.

The most important use of the theory would be to help in planning transport systems. As we showed in Chapter 12 the deterministic theory developed in this book can be used to improve the specification of the stochastic models usually used to represent and predict the number of journeys between two locations. Usually, time or distance has been used to indicate the cost of travel between two locations but our theoretical investigations suggest that in the case of the journey to work this procedure is incorrect, and the cost of travel must be measured in such a way as to allow for the systematic variation of rent and wages within the city. The cost of separation of residence and workplace varies with the distance of the place of residence from the central business area (CBD) and with the relative location of residence, workplace, and CBD. Because of the variation in rent and wages, travel towards the CBD will be considerably less costly than circumferential travel, and this, in turn, will be less expensive than travel in a direction away from the CBD.

The theory cannot only be used to help in developing models to represent the pattern of journeys to work in the city, given the pattern of location: it can also be used to predict the effect on the pattern of location of a change in the transport system. As we showed in Chapter 10 an improvement in the system will change the pattern. For example, the construction of a radial urban motorway which can be used by commuters will lead to an increase in the demand for space by high-income households on the outer boundary of the city served by the motorway. The end result will be a decentralisation of the population of the city, and of high-income households in particular, i.e. urban sprawl and a flight to the suburbs.

The analysis can also be used to assess the proposal that public transport should be regarded as a social service and provided free. It is sometimes argued that public transport should be provided free because this is the only way in which public transport can compete with private, and that, since private transport does not pay its marginal social cost, a welfare optimum which is, at least, second best can be attained. This can be seen as an alternative to road pricing, the latter ensuring that private transport pays its marginal social cost, and hopefully, that a first best optimum is attained.

The fact that implementation of either of these proposals would affect the spatial structure of cities is always neglected in discussion. Furthermore, the effects of one policy would differ completely from those of the other. Reducing the cost of public transport would allow the poor who work in the inner areas of cities to live further out. As we showed in Chapter 10 the equilibrium pattern of location in this situation is likely to be one in which the poor are dispersed away from the centre, while the rich live close to the centre. The reduction in the cost of transport is also likely to lead to a fall in the level of rents in the inner areas of the city, and an increase in the outer area coupled with an expansion in the area of the city. On the other hand, the increase in the cost of transport on introduction of road pricing would have a quite different effect. The area of the city would contract, the poor becoming more concentrated near the centre, and the rich becoming more dispersed towards the edge.

In the context of a system of cities, as analysed by Evans (1972b), we may note that, when public transport is free, the wage (and rent) premium which would have to be paid by firms locating in the largest city or cities would fall, while with road pricing it would rise. In the first case there would be an increase in the demand by firms for space in the centres of the large cities, while in the second, there would be a decrease in the demand. In the first case, therefore, the population of the largest city would tend to increase relative to the smaller cities, while in the second case, it would tend to decrease.

As we also showed in Chapter 10 a change in transport conditions is likely to lead to some areas of the city being left with housing which is at a higher density than would be optimal if the land were being redeveloped under the new conditions. In particular, a fall in the cost of transport or an increase in the speed of transport will mean that the inner area of the city is in this situation. Since, as shown in Chapter 10, this housing is unlikely to be redeveloped by private enterprise, if it is thought socially desirable to reduce densities in the area, the land would have to be redeveloped either by public agencies, as in Britain, or by subsidies to private enterprise, as in the United States (e.g. through a public agency buying the land plus buildings at one price, demolishing the buildings, and selling the land at a much lower price to a developer).

One difficulty with the British system of redevelopment by a public authority, usually the local authority, is that it is inevitably associated with the provision of subsidised housing for low-income groups and, in consequence, it is difficult to determine the extent of this subsidy. On the other hand, it is clear that if the housing of lower-income groups is to be subsidised, there is no reason why this housing should not be provided in the inner areas of cities if this is where the jobs are. It is true that, even setting aside the cost of reducing existing densities, land will be more expensive and building costs higher than if the housing were to be provided on 'green field' sites on the outer edge of the city. A subsidy given to the occupier of low-income housing in the outer suburbs may be worth less to him than the same subsidy for housing in the inner areas because of his higher journey to work costs and/or lower wages and/or lack of knowledge of the jobs available (see Kasper, 1972). The minimisation of the cost of providing public housing suggested by Stone (1959) may be mistaken because it ignores the costs imposed on the occupiers of the housing. One cannot be dogmatic on this subject, however – if employment moves to the suburbs, that is where subsidised housing for low-income households should be provided (Chapter 12). If public authorities do not respond to the change, and continue to provide subsidised housing only in the inner areas of the city, then, in turn, the subsidy given to the occupants will be reduced in value to the recipient by the necessity of travelling to the suburbs to find work, and its social benefit may be further reduced if it becomes necessary for the decentralised firms to pay higher wages than central city firms to obtain a labour force.

Note that it is not necessarily incorrect to demolish low-income housing in the city centre to construct high-income housing, as has usually been done in urban renewal in the United States, particularly at a time when the number of high-income, white-collar jobs in the centre is increasing, while the number of blue-collar, low-income jobs is decreasing. However, since filtering-down need not occur this does not mean that the former occupants of low-income housing will 'filter up' into better housing. Both high- and low-income groups live in the inner city so that either or both can be rehoused there (Chapter 8).

What is evident from the analysis in Chapter 8 of the pattern of location of residences by economic status and of possible changes in these patterns (Chapter 10) is that there is no reason why middle-income households should locate in the inner areas of cities. There are good reasons why either the rich or the poor or both will locate close to the centre, but the demand for space by middle-income groups is not low enough, and their valuation of travel time is not high enough, for their optimal location to be at the city centre. Calls for the construction of middle-income housing in the city centre, for example by Robson (1965, p. 27), are based on the fallacious assumption that the reason that middle-income groups do not live there is because there is no housing for them. If such housing were to be constructed, the subsidy per unit of space which would be required to get people to live there would be greater than would be required for high-income or low-income housing. Unless, therefore, it can be shown that the physical mixture of all income groups should be an important policy objective there seems little reason why middle-income groups should be subsidised to live in the inner city. Since there is no evidence that physical mixing encourages social mixing, and some evidence that it exacerbates social divisions (Pahl, 1970), there seems to be no reason why physical proximity of income groups should be a policy objective.

As a final illustration of the relevance of the theory of residential location to policy and planning, we shall show how the theory can be used to analyse the effects in a non-static economy, of imposing a 'green belt' around a city. We shall set out the argument in terms of the theoretical model of the city which we have used throughout this book, but the city we shall have in mind will be London. For evidence that the events predicted by the theory have occurred see, for example, Thomas (1970), Hall (1973), and Pahl (1965).

We assume that, initially, the economy of the city is in equilibrium. Each household has attained a location on the lowest attainable bid-price curve, and the rent gradient of the city is as shown by the line AA in Fig. 15.1. At this point it is enacted that no development can take place in a ring (of width AB) surrounding the built-up area of the city, because it is thought that the population of the country is going to decrease and that employment in the city is going to be reduced.

After the green belt is imposed however, employment in the

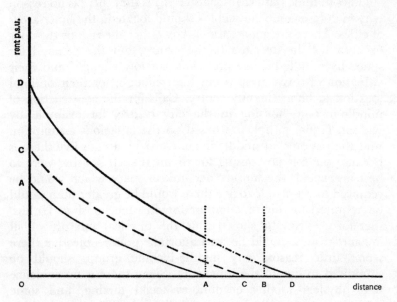

Fig. 15.1   The rent gradient with a green belt.

city continues to increase, both because the population of the country as a whole increases, and because most of the growth of employment in the economy occurs in industries which tend to be located in this large city. In the absence of a green belt the increase in the population of the city would cause the rent gradient to be bid up to the line CC in Fig. 15.1, with new development occurring in a ring of width AC, and development at increased density occurring within the circle of radius OA, Because of the green belt, however, no new development can occur in the ring of width AC and any density controls which may have been imposed over the old built-up area will prevent the increased population being accommodated by increases in density. As a result, rents and house prices will increase rapidly in the city until the rent gradient DD is attained. At this point the increased population of the city can be accommodated in new developments beyond the green belt. They will be willing to put up with the excessively long journey to work which is entailed only because of the high cost of housing inside the green belt. Similarly, those living in the old built-up area will pay the high cost of housing because the only alternative is a very long journey to work. There will, however, be pressure from those owning green-belt land, particularly land on its inner edge, for it to be 'rezoned' for development since the price of green-belt land will now be only a fraction of the price of developable land.

It may be argued in defence of the green belt that the aim is to restrict the size of this large city and that by raising land values and wages (if we assume that the high cost of housing is passed on to employers), the cost of location in the city will be increased, causing firms to move to other cities. This argument will be plausible only when no other controls are in force, or if in force, they are ineffective. Thus, in London it is clear that industrial development certificates and office development permits are the main factors limiting the increase in employment, not high wage costs.

Again, suppose that after imposition of the green belt, the transport system of the city continues to change. Increasing use of the car and electrification of railways will result in a reduction in the slope of the rent gradient, as demonstrated in Chapter 10. If the original rent gradient is shown by line AA

in Fig. 15.2, then in the absence of population growth and in the absence of the green belt, the new equilibrium-rent gradient might be *EE*. The effect of the green belt will be to prevent development in a ring of width *AE*. Rents and house prices will therefore be bid up in the outer areas of the original built-up area until a new rent gradient is attained. This might be *FF*, so that no new development takes place, or it might be *FG*, so that the transport improvement leads to development beyond the green belt. In either case, restrictions on development result in the households being on higher (i.e. less-preferred) bid-price curves than they otherwise would be.

With this analysis we can set down the costs and benefits of a green belt in a non-static economy. It is clear that the main gainers would be those owning property at the outer edge of the built-up area when the belt is imposed and those who own such property during the time taken to reach a new equilibrium. The value of their property will increase and, at the same time, they have the advantage of living near green-belt land. The main losers will be those renting property in the inner areas of the city. Their rents will increase in line with property value but they will not receive any capital gain – this will go to the landlord; furthermore, they live so far from the edge of

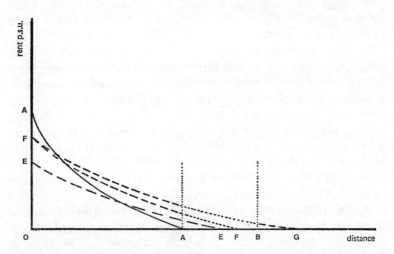

Fig. 15.2 The change in the rent gradient caused by an increase in travel speeds when there is a green belt.

the city that the existence of the green belt will scarcely affect them.†

Once the houses of owner-occupiers have changed hands at the new equilibrium prices the new owner-occupiers become losers. While the original owners made a capital gain the new owners have to pay prices inflated by the restrictions on development. Either that, or they have to make a very long journey to work. Having once paid these inflated prices, however, the new owner-occupiers have an interest in the preservation of the *status quo*, since they would stand to lose all the capital gains received by the previous owners. Thus, any relaxation in the application of green-belt policy would have to take place gradually in order not to cause a politically disruptive fall in housing prices. Because rents and house prices in the old built-up area of the city are higher than they would be if the green belt did not exist, it follows that, when sites within this area are redeveloped, they will be redeveloped at densities which are higher than they otherwise would be.

Whether this is good or bad is open to discussion – there are external economies and diseconomies of high density. On the one hand, the higher densities allow, for example, a better public transport system and greater proximity to shops and services, while on the other, lower densities mean that the city has a more open appearance and there is a greater use of private transport. What is clear is that if there is any relaxation in the application of green-belt policy so that rents and house prices fall, then, as we showed in Chapter 10, the high-density developments are likely to remain long after their 'normal' economic life.

To summarise, the policy imposes costs on all residents in the city which probably outweigh the possible benefits. The main benefits accrue to the well-off living in their own houses in the suburbs, while the main costs are borne by the poor in rented housing near the city centre. Thus, the green belt fails both of the two tests of a microeconomic policy, it is both inefficient and inequitable. It is also probable that the increase in house prices generated by restrictions on development may

† Their access to open space may even be reduced if the green belt causes the local authority to solve its housing problem by building over open space lying within the boundary of the old built up area.

fuel a more general inflation. Yet, once a green belt has been imposed it becomes politically difficult to rezone land on its inner edge for redevelopment, because of the opportunities of speculation and the vast capital gains which would accrue to the landowners. In Britain this problem would have to be solved by using a form of New Town Development Corporation for the development of suburbs instead of satellite towns.

We still lack a normative theory of residential location, and I hope that the development of the positive theory in this book will stimulate its development. Yet the above discussion of the economics of the green belt, and the discussion of various policies earlier in the chapter, clearly indicate that the positive theory of residential location can still be useful to those drawing up plans or advocating policies for the city.

Appendix A

Muth (1969, p. 21) writes the utility function

$$U = U(x, q) \qquad \text{(A.1)}$$

where q is consumption of housing, and x is expenditure on all commodities except housing and transportation but including leisure. The only constraint on the consumer's attempt to maximise utility is a budget constraint, which Muth writes:

$$x + p(k) \cdot q + T(k, y) - y = 0 \qquad \text{(A.2)}$$

where p, the price per unit of housing, is a function of k, distance from the CBD. T is the total cost of journeys to the CBD and is the sum of the direct financial cost of trips to the CBD and of the indirect cost, which is the imputed value of travel time. T is a function of k, distance from the CBD, and also of y, income, since an increase in income increases the value of travel time. The first-order conditions which Muth states to be necessary for the maximisation of A.1, subject to A.2, are (Muth, 1969, p. 22):

$$U_x - \lambda = 0$$
$$U_q - \lambda \cdot p = 0$$
$$-\lambda (q \cdot p_k + T_k) = 0 \qquad \text{(A.3)}$$

and

$$y = [x + p(k) \cdot q + T(k, y)] = 0$$

where λ is a Lagrange multiplier. The only condition which is relevant for the theory of residential location is the third (A.3), and we will show that, given Muth's assumptions, the condition is incorrect. In the budget constraint (A.2), income, y, is written as though it were a constant, as of course, it would be if y were equal to the wage and property income of the household for the period. But in his discussion of time costs, Muth states that 'I shall simply assume that time costs are a function of the wage rate; . . . *I include travel time costs in trip costs and the money value of travel time in the income variable in the consumer's budget constraint*' (p. 20, my italics).

Now if the money value of travel time is included in the income variable, it follows that y is not a constant but a function of distance from the CBD, k; for the longer the time spent in travelling, the higher is the money value of travel time, and the higher is y. Therefore, $y = y(k)$ in the budget constraint (A.2) and the third condition should be written:

$$-\lambda(q.p_k + T_k - y_k) = 0. \qquad (A.4)$$

This expression can be simplified, however. Let $T(k, y) = c(k) + t(k, y)$, where $c(k)$ is the direct financial cost of travel and $t(k, y)$ is the value of travel time, and let $y = \bar{y} + t(k, y)$, where $\bar{y}$ is the household's money income. The budget constraint can then be written:

$$x + p(k).q + c(k) + t(k, y) - \bar{y} - t(k, y) = 0. \qquad (A.5)$$

Note that this is exactly the same constraint as (A.2); we have merely expanded two terms. If we maximise (A.1) subject to (A.5) we obtain, instead of (A.3),

$$-\lambda(q.p_k + c_k) = 0 \qquad (A.6)$$

It is clear that, since the value of travel time is added both to money travel costs and to money income, it is cancelled out and the equilibrium conditions are the same as if the value of time was never considered at all. Since the valuation of time is important, and does affect the consumers location decision, it is clear that Muth's method of incorporating the valuation of time into his analysis is incorrect. It may be noted that Solow (1972) uses the same method in an analysis of congestion, density, and land use, and the same criticism applies.

Appendix B

Richardson (1971) rejects the trade-off theory of residential location and puts forward suggestions for a 'behavioural theory' of residential location. He argues that, for owner occupiers, 'housing preferences (including the kind of area and quality of environment the household desires) and financial constraints, that is income and the conditions and availability of mortgage finance, are the primary independent variables, and that journey to work costs are at best a secondary determinant . . . [although] the journey to work must act as a constraint in that there is a maximum commuting limit in travel time (p. 24f.).

Many of his criticisms of trade-off theories are implicitly answered in this book, in which the trade-off theory is used to explain and predict actual patterns of location, e.g. the location of the rich at the centre. Furthermore, as stated in Chapter 2, although we do not explicitly deal with the environment we do incorporate social agglomeration into the theory, and we do consider the influence of the natural environment on the pattern of location. Both these factors would seem to be important determinants of the characteristics of an area and the quality of the environment.

Furthermore, a number of criticisms can be made of Richardson's own theory. Firstly Richardson assumes the rent gradient to be given exogenously, rather than determined by the demand for space of the households in the city, as it is in the trade-off theory. This means, presumably, that the rent gradient must be determined by the demands for space of industrial and commercial firms, and this seems an unrealistic assumption since they occupy a far smaller proportion of the area of the city than households do. Secondly, Richardson states that 'in most cases the journey to work costs will not have much of an influence provided that the area of search falls within commuting limits' (p. 25). But the financial costs of commuting presumably do matter, and many studies have shown that workers put a value on commuting time which does not vary much with distance. It therefore seems logical to assume that journey to work costs do have an influence on location.

Finally, although it is quite possible that the conditions and availability of mortgage finance limit the amount which an owner-occupier can spend on the purchase of a house, there is no evidence that this constraint is binding. Thus, Byatt, Holmans, and Laidler (1972), after a study of British building society data which they wished to use to estimate the income elasticity of demand for housing, concluded that 'households tend to borrow less in relation to income where prices are relatively low'. They also showed that the availability of building society funds had 'only a very minor influence on the measured income elasticity', and thought that it was 'probably safe to conclude that data on housing transactions drawn from Building Society sources does show something about the economic behaviour of house purchasers, and not merely about Building Society practice' (p. 9). This would not be true if a capital constraint were usually effective.

Nevertheless, the idea that an important factor determining the location of a household is the amount secured by a mortgage which it can borrow can be incorporated into the trade-off theory. Two assumptions are necessary: first, households wish to own their own houses rather than rent them, possibly because of favourable tax treatment for owner-occupiers,† and second, the institutions which finance house purchase (in the U.K., mainly building societies) are not willing to lend more than a certain multiple of the household's annual income. As a result, the household cannot purchase a house costing more than a certain amount, and if it wishes to occupy a property which would cost more, it must rent it, even though this would be more expensive for the household.

The effects of this capital constraint on the household's

† In the U.K. mortgage interest is allowable against tax, while rent is not. In addition, even when the purchase of the property is wholly financed by the owner, the notional income from his investment in his home (i.e. the rent which, as owner, he might charge himself as occupier) is not chargeable to tax.

choice of location can be shown diagrammatically. In Fig. B.1 price per space unit (i.e. either the rent or its capitalised value) is shown on the vertical axis, and distance from the CBD is shown on the horizontal axis. The line $ABCD$ is a representative bid-price curve; it has a discontinuous slope because of the capital constraint. If the householder were always to rent space, then, for a given utility level U_0, the bid-price curve would be the line ABE. If, on the other hand, he was always to buy space, and the capital constraint was inoperative, then, because of tax concessions, he could afford to pay a higher price for space at a given location than if he were to rent it, but would still attain the same utility level. Hence, for the given utility level U_0, the bid-price curve if he were always to

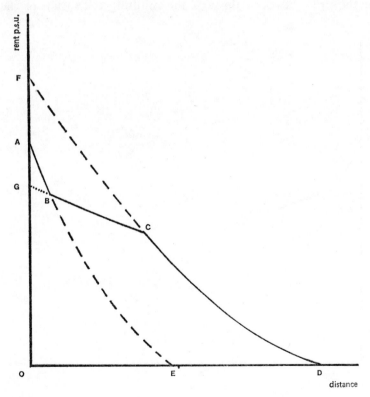

Fig. B.1   A bid-price curve when there is a capital constraint.

buy space would be the line *FCD*, provided there is no capital constraint.

If, however, the householder's demand for space is relatively price inelastic, so that the amount he would spend on space in the inner areas of the city would be higher than the amount he would spend in the outer areas, the capital constraint may prevent him from buying all the space he would wish to buy in the inner city. Points on the line *FC* will therefore be un-attainable. A third bid-price curve, *GBC*, must be drawn for the given utility level U_0, indicating the prices the household is willing to pay to buy space in the inner city when he is unable to spend more than a fixed amount. This line still slopes downwards but is less steep than *FCD* because the householder gains from each move outward and the consequent reduction in the price of space through the resulting relaxation of the

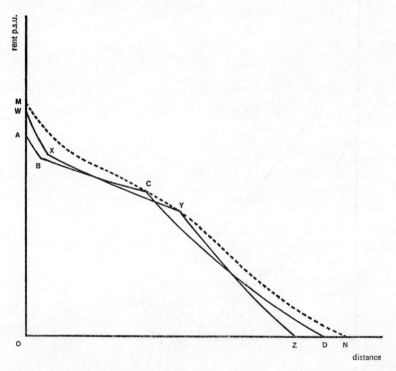

Fig. B.2   Equilibrium when there are capital constraints.

capital constraint. The representative bid-price curve is thus made up of sections *AB*, *BC*, and *CD* of the three other curves.

Some possible effects of the capital constraint in these circumstances are shown in Fig. B.2. Because of the discontinuity in the slope of the bid-price curve at the point at which the capital constraint becomes effective, and the consequent non-convexity of the bid-price curve, this is the most likely point of tangency between the rent gradient and the lowest attainable bid-price curve, as shown at points *C* and *Y* in Fig. B.2. As a result the rent gradient will be 'distorted' and may become non-convex at intermediate distances from the centre, as shown in Fig. B.2 by the line *MCYN*. A further result is that a high-income household may find its optimal location closer to the centre than a household with a lower-income will, even though the relative slopes of the sets of bid-price curves of the two households would indicate that the former should live further out. This is shown in Fig. B.2, in which each section of the bid-price curve *ABCD* is less steep than the corresponding part of the bid-price curve *WXYZ*, but the higher-income household's optimal location is indicated by *C*, which is closer to the centre than *Y*, thus indicating the optimal location of the lower-income household.

A capital constraint can affect the location pattern in other ways. For example, if the demand for space of a household is relatively price elastic the capital constraint will become effective *beyond* a certain distance from the centre, when the price falls low enough. As a result, a low-income household may find its optimal location closer to the centre than a high-income household will, even though its bid-price curves are less steep.

References

Abercrombie, Patrick (1959). *Town and Country Planning*, ed. D. Rigby Childs, 3rd ed. (London: Oxford University Press).

Abu-Lughod, Janet (1960). 'A Survey of Center-city Residents', in Nelson N. Foote, et al., *Housing Choices and Housing Constraints* (New York: McGraw-Hill).

Abu-Lughod, Janet (1969) 'Testing the Theory of Social Area Analysis: the Ecology of Cairo, Egypt', *American Sociological Review*, 34 (Apr).

Alonso, William (1960) 'A Theory of the Urban Land Market', *Papers and Proceedings of the Regional Science Association*, 6.

Alonso, William (1964a) *Location and Land Use: Toward a General Theory of Land Rent* (Cambridge, Mass: Harvard University Press).

Alonso, William (1964b) 'The Historic and the Structural Theories of Urban Form: their Implications for Urban Renewal', *Land Economics*, 40 (May).

Anderson, Theodore R. (1962) 'Social and Economic Factors Affecting the Location of Residential Neighbourhoods', *Papers and Proceedings of the Regional Science Association*, 9.

Anderson, Theodore R, and Egeland, Janice (1961) 'Spatial Aspects of Social Area Analysis', *American Sociological Review*, 26 (June).

Beckmann, Martin (1969) 'On the Distribution of Urban Rent and Residential Density', *Journal of Economic Theory*, 1 (June).

Beed, Clive (1970) 'The Development of the Least Cost Theory of Residential Location', in *The Analysis of Urban Development*, ed. Nicholas Clark (Melbourne: Department of Civil Engineering, University of Melbourne).

Berry, Brian J. L., and Horton, F. E. (1970) *Geographical Perspectives on Urban Systems* (Englewood Cliffs, N.J.: Prentice-Hall).

Berry, Brian J. L., Simmons, James W., and Tennant, Robert J. (1963) 'Urban Population Densities: Structures and Change', *The Geographical Review*, 53 (July).

Boyce, David E. (1965) 'The Effect of Direction and Length of Person Trips on Urban Travel Patterns, *Journal of Regional Science*, 6 (Summer).

Brigham, Eugene F. (1965) 'The Determinants of Residential Land Values', *Land Economics*, 41 (Nov).

Buchanan, James M. (1965) 'An Economic Theory of Clubs', *Economica*, 32 (Feb).

Burgess, Ernest W. (1925) 'The Growth of the City: an Introduction to a Research Project', in *The City*, ed. R. E. Park, E. W. Burgess, and R. A. McKenzie (Chicago: University of Chicago Press).

Buttimer, Anne (Sister Mary Annette, O.P.) (1969) 'Social Space in Inter-disciplinary Perspective', *The Geographical Review*, 59 (July).

Byatt, I. C. R., Holmans, A. E., and Laidler, D. E. W. (1972) 'Income and the Demand for Housing; some Evidence for Great Britain', (paper given to the conference of the Association of University Teachers of Economics at Aberystwyth).

Carroll, J. Douglas, Jr. (1952) 'The Relation of Homes to Work Places and the Spatial Pattern of Cities', *Social Forces*, 30 (Mar).

Casetti, E. (1969) 'Alternate Urban Density Models: an Analytical Comparison of their Validity Range', in *Studies in Regional Science*, ed. Allen J. Scott (London: Pion).

Clark, Colin (1951) 'Urban Population Densities', *Journal of the Royal Statistical Society*, ser. A 114 no. 4.

Clark, Colin (1966) 'Urban Land Use Here and Abroad', *Journal of the Town Planning Institute*, 52 (Nov).

Clark, Colin (1967) *Population Growth and Land Use* (London: Macmillan).

Clark, Colin, and Peters, G. H. (1965) 'The Intervening Opportunities Method of Traffic Analysis', *Traffic Quarterly*, 19 (Jan.).

Craven, Edward (1969) 'Private Residential Expansion in Kent 1956–64: a Study of Pattern and Process in Urban Growth', *Urban Studies*, 6 (Feb).

Crecine, John P., Davis, Otto A., and Jackson, John E (1967) 'Urban Property Markets: some Empirical Results and their Implications for Municipal Zoning', *Journal of Law and Economics*, 10 (Oct).

Dalvi, M. Q., and Lee, N. (1971) 'Variations in the Value of Travel Time: Further Analysis', *Manchester School of Economic and Social Studies*, 39 (Sept).

Davis, Otto A., and Whinston, Andrew B. (1961) 'The Economics of Urban Renewal', *Law and Contemporary Problems*, 26 (winter).

De Leeuw, Frank (1971) 'The Demand for Housing: a Review of Cross-section Evidence', *Review of Economics and Statistics*, 53 (Feb).

DeSerpa, A. C. (1971) 'A Theory of the Economics of Time', *Economic Journal*, 81 (Dec).

Detroit Metropolitan Area Transportation Study (1953) *Data Summary and Interpretation* (Detroit).

Devetoglou, N. E. (1965) 'A Dissenting View of Duopoly and Spatial Competition', *Economica*, 32 (May).

Dorfman, Robert (1969) 'An Economic Interpretation of Optimal Control Theory', *American Economic Review*, 59 (Dec).

Evans, Alan W. (1972a) 'On the Theory of the Valuation and Allocation of Time', *Scottish Journal of Political Economy*, 19 (Feb).

Evans, Alan W. (1972b) 'The Pure Theory of City Size in an Industrial Economy', *Urban Studies*, 9 (Feb).

Evans, Alan W. (1972c) 'Residential Location in Cities' (Ph.D. Thesis, University of London).

Evely, R., and Little, I. M. D. (1960) *Concentration in British Industry* (Cambridge: Cambridge University Press).

Fagin, Henry (1967) 'Sprawl and Planning', in *Metropolis on the Move*, ed. Jean Gottmann and Robert A. Harper (New York: John Wiley).

Frayn, Michael (1967) *Towards the End of the Morning* (London: Collins).

Friedman, M. (1953) 'The Methodology of Positive Economics', in *Essays in Positive Economics* (Chicago: University of Chicago Press).

Frieden, Bernard J. (1961) 'Locational Preferences in the Urban Housing Market', *Journal of the American Institute of Planners*, 27 (Nov).

Fuchs, Victor R. (1967) *Differentials in Hourly Earnings by Region and City Size, 1959*, N.B.E.R. Occasional Paper 101 (New York: Columbia University Press, for National Bureau of Economic Research).

Gittus, Elisabeth (1965) 'An Experiment in the Definition of Urban Sub-areas', *Transactions of the Bartlett Society*.

Granelle, J-J. (1968) 'La Formation des Prix du Sol dans l'Espace Urbain', *Revue D'Economie Politique*, 78 (Jan).

Grebler, Leo, Blank, David M., and Winnick, Louis (1956) *Capital Formation in Residential Real Estate* (Princeton: Princeton University Press).

Green, L. P. (1959) *Provincial Metropolis* (London: George Allen & Unwin).

Grigsby, William G. (1963) *Housing Markets and Public Policy* (Philadelphia: Pennsylvania University Press).

Haig, R. M. (1926) 'Toward an Understanding of the Metropolis, I and II', *Quarterly Journal of Economics*, 40 (Feb. and May).

Hall, Peter (1962) *The Industries of London* (London: Hutchinson).

Hall, Peter, et. al (1973) *The Containment of Urban England* (London: George Allen & Unwin).

Harris, Britton (1968) 'Quantitative Models of Urban Development', *Issues in Urban Economics*, ed. Harvey S. Perloff and Lowdon Wingo, Jr. (Baltimore: The Johns Hopkins Press for Resources for the Future).

Harris, Chauncy D., and Ullman, Edward L. (1945) 'The Nature of Cities,' *Annals of the American Academy of Political and Social Science*, 242 (Nov).

Hawley, Amos (1955) 'Land Value Patterns in Okayama, Japan, 1940 and 1952', *American Journal of Sociology*, 60 (Mar).

Henderson, J. M., and Quandt, R. E. (1958) *Microeconomic Theory* (New York: McGraw Hill).

Herbert, D. T. (1967) 'Social Area Analysis: a British study', *Urban Studies*, 4 (Feb).

Herbert, D. T. (1970) 'Principal Components Analysis and Urban Social Structure: a Study of Cardiff and Swansea', *Urban Essays: Studies in the Geography of Wales*, ed. H. Carter and W. K. D. Davies (London: Longman).

Hicks, John (1969) *A Theory of Economic History* (London: Oxford University Press).

Hoch, Irving (1969) 'The Three Dimensional City: Contained Urban Space', *The Quality of the Urban Environment*, ed. Harvey S. Perloff (Washington, D.C.: Resources for the Future).

Hoch, Irving (1972) 'Income and City Size', *Urban Studies*, 9 (Oct).

Hoover, Edgar M. (1948) *The Location of Economic Activity* (New York: McGraw Hill).

Hoover, Edgar M., and Vernon, Raymond (1959) *Anatomy of a Metropolis* (Cambridge, Mass: Harvard University Press).

Hoyt, Homer (1939) *The Structure and Growth of Residential Neighbourhoods in American Cities* (Washington: U.S. Government Printing Office; repr. Homer Hoyt Associates, 1968).

Institute of Office Management (1962) *Clerical Salaries Analysis: 1962* (London: Institute of Office Management [now Institute of Administrative Management]).

Isard, W. (1956) *Location and Space-Economy* (Cambridge, Mass.: M.I.T. Press).

Jacobs, Jane (1961) *The Death and Life of Great American Cities* (New York: Random House).

Johnston, R. J. (1969) 'Some Tests of a Model of Intra-Urban Population Mobility: Melbourne, Australia', *Urban Studies*, 6 (Feb).

Johnston, R. J. (1971) *Urban Residential Patterns* (London: Bell).

Kain, John F. (1962) 'The Journey to Work as a Determinant of Residential Location', *Papers and Proceedings of the Regional Science Association*, 9.

Kain, John F. (1968) 'Housing Segregation, Negro Employment and Metropolitan Decentralization', *Quarterly Journal of Economics*, 82 (May).

Kasper, Hirschel (1972) 'Measuring the Labour Market Costs of Housing Dislocation', Urban and Regional Studies Discussion Paper no. 4, (Dept. of Social and Economic Research, University of Glasgow).

Kirwan, Richard (1966) Review of *Location and Land Use* by William Alonso *Urban Studies*, 3 (June).

Knos, Duane S. (1968) 'The Distribution of Land Values in Topeka, Kansas', *Spatial Analysis*, ed. B. J. L. Berry and D. F. Marble (Englewood Cliffs, N.J.: Prentice-Hall).

Lancaster, Kelvin (1968) *Mathematical Economics* (New York: Macmillan & Co.).

Lane, Robert (1970) 'Some Findings on Residential Location, House Prices, and Accessibility' (paper given to the Research and Intelligence Unit, Department of Planning and Transportation, Greater London Council).

Lawton, R. (1968) 'The Journey to Work in Britain: Some Trends and Problems', *Regional Studies*, 2 (Sep).

Lee, N., and Dalvi, M. Q. (1969) 'Variations in the Value of Travel Time', *Manchester School of Economic and Social Studies*, 37 (Sep).

Lester, R. A. (1967) 'Pay Differentials by Size of Establishment', *Industrial Relations*, 7 (Oct).

Lever, W. F. (1971) 'Planning Standards and Residential Densities', *Journal of the Royal Town Planning Institute*, 57 (Nov).

Lipman, V. D. (1954) *A Social History of the Jews in England, 1850–1950* (London: Watts).

Liepmann, Kate (1944) *The Journey to Work* (London: Kegan Paul).

London Traffic Survey (1964) Vol. 1 (London County Council).

Lösch, August (1954) *The Economics of Location*, trans. W. H. Woglom (New Haven: Yale University Press).

Lowry, Ira S. (1960) 'Filtering and Housing Standards', *Land Economics*, 36 (Nov.).

McDonald, Ian J. (1969) 'The Leasehold System: towards a Balanced Land Tenure for Urban Development', *Urban Studies*, 6 (June).

McElrath, Dennis (1962) 'The Social Areas of Rome: a Comparative Analysis', *American Sociological Review*, 27 (June).

Mackay, D. I., Boddy, D., Brack, J., Diack, J. A., and Jones, N. (1971) *Labour Markets under Different Employment Conditions* (London: George Allen and Unwin).

Malamud, B. (1971) 'The Economics of Office Location'. (Ph.D. thesis, New School for Social Research).

Mansfield, Edwin (1957) 'City Size and Income, 1949', *Regional Income* (Princeton University Press, for National Bureau of Economic Research).

Martin, J. E. (1969) 'Size of Plant and Location of Industry in Greater London', *Tijdschrift voor Economische en Sociale Geografie*, 40 (Nov/Dec).

Mattila, John M, and Thompson, Wilbur R. (1968) 'Appendix: Toward an Econometric Model of Urban Economic Development', in *Issues in Urban Economics*, ed. Harvey S. Perloff and Lowdon Wingo Jr. (Baltimore: The Johns Hopkins Press, for Resources for the Future).

Medhurst, D. F., and Lewis, J. Parry (1969) *Urban Decay* (London: Macmillan).

Mills, Edwin S. (1969) 'The Value of Urban Land', in *The Quality of the Urban Environment*, ed. Harvey S. Perloff (Washington, D.C.: Resources for the Future).

Mills, Edwin S. (1970) 'Urban Density Functions', *Urban Studies*, 7 (Feb).

Ministry of Housing and Local Government (1965) *Report of the Committee on Housing in Greater London*, chaired by Sir Milner Holland (London: H.M.S.O.).

Ministry of Housing and Local Government (1966) *The Housing Cost Yardstick* (London: H.M.S.O.).

Ministry of Labour (1959) 'Average Earnings and Hours of Men in Manufacturing: Analysis by Size of Establishment', *Ministry of Labour Gazette*, 67 (Apr).

Ministry of Transport (1969) 'The Value of Time Savings in Transport Investment Appraisal' (a paper prepared by the Economic Planning Directorate, London).

Mirrlees, James A. (1972) 'The Optimum Town', *Swedish Journal of Economics*, 74 (Mar).

Mishan, E. J. (1961) 'Theories of Consumer's Behaviour: a Cynical View', *Economica*, 28 (Feb).

Mitchell, Robert and Rapkin, Chester (1954) *Urban Traffic: a Function of Land Use* (New York: Columbia University Press).

Moses, Leon N. (1962) 'Towards a Theory of Intra-Urban Wage Differentials and their Influence on Travel Patterns', *Papers and Proceedings of the Regional Science Association*, 9.

Murdie, Robert A. (1969) *Factorial Ecology of Metropolitan Toronto, 1951–1961*, Department of Geography Research Paper no. 116 (Chicago: University of Chicago).

Muth, R. F. (1969) *Cities and Housing* (Chicago: University of Chicago Press).

Needleman, Lionel (1969) 'The Comparative Economics of Improvement and New Building', *Urban Studies*, 6 (June).

Neutze, Max (1968) *The Suburban Apartment Boom* (Washington, D.C.: Resources for the Future).

Newling, Bruce E. (1969) 'The Spatial Variation of Urban Population Densities', *The Geographical Review*, 59 (Apr).

Pahl, R. E. (1965) *Urbs in Rure – The Metropolitan Fringe in Hertfordshire* (London: L.S.E. Geographical Papers, no. 2).

Pahl, R. E. (1970) *Patterns of Urban Life* (London: Longmans, Green).

Pap, Arthur (1963) *The Philosophy of Science* (London: The Free Press).

Pederson, Poul Ove (1967) 'An Empirical Model of Urban Population Structure: a Factor Analytic Study of the Population Structure of Copenhagen', *Proceedings of the First Scandinavian – Polish Regional Science Seminar*, (Warsaw: Polish Scientific Publishers).

Rees, Albert and Shultz, George P. (1970) *Workers and Wages in an Urban Labor Market* (Chicago: Chicago University Press).

Rees, Philip H. (1970a) 'The Axioms of Intra-Urban Structure and Growth' [part of Chapter 9 of Berry and Horton (1970)].

Rees, Philip H. (1970b) 'Concepts of Social Space: Toward an Urban Social Geography' [Chapter 10 of Berry and Horton (1970)].

Reid, Margaret G. (1962) *Housing and Income* (Chicago: University of Chicago Press).

Richardson, Harry (1971) *Urban Economics* (London: Penguin Education).

Robinson, Joan (1965) 'The Production Function and the Theory of Capital', *Collected Economic Papers* (Oxford: Basil Blackwell).

Robson, Brian T. (1969) *Urban Analysis: a Study of City Structure with Special Reference to Sunderland* (Cambridge: Cambridge University Press).

Robson, William A. (1965) *The Heart of Greater London: Proposals for a Policy*, Greater London Papers, no 9 (London: London School of Economics and Political Science).

Rossi, Peter H. (1955) *Why Families Move* (New York: The Free Press).

Rummel, R. J. (1967) 'Understanding Factor Analysis', *Journal of Conflict Resolution*, 11 (Dec).

Schnore, Leo F. (1954) 'The Separation of Home and Work: a Problem for Human Ecology', *Social Forces*, 32 (May).

Schnore, Leo F. (1965) 'On the Spatial Structure of Cities in the Two Americas', in *The Study of Urbanization*, ed. Philip M. Hauser and Leo F. Schnore (New York: John Wiley).

Segal, Martin (1960) *Wages in the Metropolis* (Cambridge, Mass.: Harvard University Press).

Seyfried, Warren R. (1963) 'The Centrality of Urban Land Values', *Land Economics*, 39 (Aug).

Shevky, Eshref, and Bell, Wendell (1955) *Social Area Analysis* (Stanford: Stanford University Press).

Siegel, Sidney (1956) *Nonparametric Statistics for the Behavioral Sciences* (New York: McGraw Hill).

Simmons, James W. (1968) 'Changing Residence in the City: a Review of Intra-Urban Mobility', *The Geographical Review*, 58 (Oct). [Reprinted in Chapter 11 of Berry and Horton (1970)].

Sjoberg, Gideon (1965) 'Cities in Developing and in Industrial Societies: a Cross-cultural Analysis, in *The Study of Urbanization*, ed. Philip M. Hauser and Leo F. Schnore (New York: John Wiley).

Solow, Robert M. (1972) 'Congestion, Density, and the Use of Land in Transportation', *Swedish Journal of Economics*, 74 (Mar).

Stone, P. A. (1959) 'The Economics of Housing and Urban Development', *Journal of the Royal Statistical Society*, ser. A 122 no. 4).

Stone, P. A. (1964) 'The Price of Sites for Residential Building', *The Estates Gazette*, 189 (11 Jan).

Stone, P. A. (1965) 'The Prices of Building Sites in Britain', in *Land Values*, ed. Peter Hall (London: Sweet & Maxwell).

Sweetser, Frank L. (1961) *The Social Ecology of Metropolitan Boston: 1950* (Boston: Massachusetts Department of Mental Health).

Sweetser, Frank L. (1962) *The Social Ecology of Metropolitan Boston: 1960* (Boston: Massachusetts Department of Mental Health).

Taaffe, Edward J., Garner, Barry J., and Yeates, Maurice H. (1963) *The Peripheral Journey to Work* (Chicago: Northwestern University Press).

Thomas, David, (1970) *London's Green Belt* (London: Faber & Faber).

Timms, D. W. G. (1971) *The Urban Mosaic: towards a Theory of Residential Differentiation* (London: Cambridge University Press).

Tindall, Gillian (1971) 'A Street in London', *New Society*, 17 (14 Jan).

Turvey, Ralph (1957) *The Economics of Real Property* (London: George Allen & Unwin).

Turvey, Ralph (1964) Review of *Location and Land Use* by William Alonso, *American Economic Review*, 54 (Sep).

Wabe, J. S. (1969) 'Commuter Travel into Central London', *Journal of Transport Economics and Policy*, 3 (Jan).

Wabe, J. S. (1971) 'A Study of House Prices as a means of Establishing the Value of Journey Time, the Rate of Time Preference, and the Valuation of some Aspects of Environment in the London Metropolitan Region', *Applied Economics*, 3 (Dec).

Warnes, A. M. (1968) 'Changing Journey to Work Patterns: an indicator of metropolitan change' (paper presented to the I.B.G. Study Group in Urban Geography at Salford, Lancashire).

Weingartner, H. Martin, (1969) 'Some New Views on the Payback Period and Capital Budgeting Decisions', *Management Science: Application Series*, 15 (Aug).

Westergaard, John (1957) 'Journeys to Work in the London Region', *Town Planning Review*, 28 (Apr).

Widmer, Georges (1953) L'Inegalité dans la Grandeur des Villes et ses Correlations Economiques, *Revue Economique*, 4 (May).

Wilkinson, R., and Merry, D. M. (1965) 'Statistical Analysis of Attitudes to Moving', *Urban Studies*, 2 (May).

Willmott, Peter, and Young, Michael (1960) *Family and Class in a London Suburb* (London: Routledge & Kegan Paul).

Wilson, A. G. (1967) 'A Statistical Theory of Spatial Distribution Models', *Transportation Research*, 1, no. 3.

Wilson, F. R. (1968) *Journey to Work* (London: Maclaren).

Wingo, Lowdon, Jr (1961a) *Transportation and Urban Land* (Washington, D.C.: Resources for the Future).

Wingo, Lowdon Jr. (1961b) 'An Economic Model of the Utilisation of Urban Land for Residential Purposes', *Papers and Proceedings of the Regional Science Association*, 7.

Yeates, Maurice H. (1965) 'Some Factors Affecting the Spatial Distribution of Chicago Land Values, 1910–60', *Economic Geography*, 41 (Jan).

Index